T0188663

ANNALS *of* THE NEW YORK ACADEMY OF SCIENCES

EDITOR-IN-CHIEF
Douglas Braaten

ASSOCIATE EDITOR
Rebecca E. Cooney

PROJECT MANAGER
Steven E. Bohall

EDITORIAL ADMINISTRATOR
Daniel J. Becker

Artwork and design by Ash Ayman Shairzay

The New York Academy of Sciences
7 World Trade Center
250 Greenwich Street, 40th Floor
New York, NY 10007-2157

annals@nyas.org
www.nyas.org/annals

**The New York
Academy of Sciences**

Published by Blackwell Publishing
On behalf of the New York Academy of Sciences

Boston, Massachusetts
2012

ANNALS *of* THE NEW YORK ACADEMY OF SCIENCES

VOLUME
1262

ISSUE

Neuroimmunomodulation in Health and Disease II

Translational Science

ISSUE EDITORS

Adriana del Rey,[a] C. Jane Welsh,[b] Markus J. Schwarz,[c] and Hugo O. Besedovsky[a]

[a]University of Marburg, [b]Texas A&M University, and [c]Ludwig-Maximilian University

TABLE OF CONTENTS

Neuroimmunomodulation and neuropsychiatric diseases

Neuroimmunomodulation of pain, and neoplastic diseases

Ann. N.Y. Acad. Sci. ISSN 0077-8923

ANNALS OF THE NEW YORK ACADEMY OF SCIENCES
Issue: *Neuroimmunomodulation in Health and Disease*

Foreword for *Neuroimmunomodulation in Health and Disease*

The immune system contributes to maintaining constancy of molecular and cellular components in an organism, and to facilitating adaptation not only to changeable environmental but also endogenous conditions. As in all physiological systems, the immune system is under the control of the neuroendocrine system and it interacts and works in concerted ways with other systems of the body. These interactions are based on the capacity of immune cells to produce chemical mediators—such as cytokines—capable of affecting the nervous and endocrine systems. Immune cells respond to neuroendocrine signals by way of specific receptors for hormones, neurotransmitters, and neuropeptides expressed on immune cells at different stages of their development and activation. This results in a neuroendocrine level of regulation that is interwoven with autoregulatory immune mechanisms. On the other hand, some immune products—for example, certain cytokines—are produced in the brain under basal conditions and, following increased neuronal and immune activity, contribute to modulate brain function. Furthermore, under particular conditions, immune, neural, and endocrine cells produce the same mediators and share the use of second messengers, protein kinases, transcription factors, and other intracellular mediators, and regulate posttranscriptional events. Since the immune, endocrine, and nervous systems are always active and their signals are produced at multiple cellular and organ levels, terms such as *bidirectional communication, feed-back,* and *reflexes* have become too restricted; we now realize that neuro–endocrine–immune interactions can only be considered as dynamic networks that tend to reach an adaptive equilibrium during health and disease. The growing evidence that stimuli that originally were thought only to elicit an immune response can also trigger neural and endocrine responses, and that those that were thought only to influence the brain and associated mechanisms can also exert immunoregulatory effects, indicates the expanded operation of this network. This network of interactions also mediates metabolic adjustments that can influence both immune and brain functions. As with all physiological mechanisms, disruptions, distortions, or imbalances in the network may lead to disease.

This and an accompanying volume of *Annals of the New York Academy of Sciences* present contributions on the main topics discussed at the 8th Congress of the International Society for Neuroimmunomodulation (ISNIM), held on October 20–22, 2011, and organized with the German Endocrine-Brain-Immune Network in Dresden, Germany. The oral presentations at the congress were selected to provide a wide spectrum of research devoted to interactions among the immune, endocrine, and nervous systems. As indicated by the name chosen for this congress, Neuroimmunomodulation in Health and Disease: An Integrative Biomedical Approach, speakers discussed not only basic research regarding neuro–endocrine–immune interactions but also the potential relevance of research for clinical practice. Consistent with this aim, the contributions

doi: 10.1111/j.1749-6632.2012.06678.x

Ann. N.Y. Acad. Sci. 1262 (2012) vii–viii © 2012 New York Academy of Sciences.

presented in both *Annals* volumes have been divided into the basic and translational aspects of neuroimmunomodulation.

The first volume, *Neuroimmunomodulation in Health and Disease I: Basic Science*, covers selected molecular mechanisms underlying neuro–endocrine–immune interactions, the relevance of these interactions for the development of the immune system and thymic function, and the establishment of self-tolerance. Examples of papers in this section include those describing how immune cells can influence the innervation of lymphoid organs and how neurotransmitters and hormones affect immune and inflammatory processes. The possible implications of complex neuroendocrine responses for immunoregulation during stress, sleep, and aging, and also during the course of the body's circadian rhythms, are also included in this section.

The second volume, *Neuroimmunomodulation in Health and Disease II: Translational Science*, includes studies dealing with infectious diseases, such as tuberculosis, HIV, and parasitic diseases. These studies are followed by reports related to neuropsychiatric disorders such as schizophrenia, depression, and Alzheimer's disease. This volume also presents examples of neuro–immune–endocrine interactions during pain, and in autoimmune, inflammatory, and neoplasic diseases.

We thank the authors for their valuable contributions and the reviewers for their helpful comments. And while we regret that not all of the excellent oral and poster presentations at the congress could be included in these volumes, we trust that the selected studies are representative examples of the intense efforts currently being made to provide further relevance for integrative views of biology, physiology, and medicine.

ADRIANA DEL REY
University of Marburg

C. JANE WELSH
Texas A&M University

MARKUS J. SCHWARZ
Ludwig-Maximilian University

HUGO O. BESEDOVSKY
University of Marburg

Ann. N.Y. Acad. Sci. ISSN 0077-8923

ANNALS OF THE NEW YORK ACADEMY OF SCIENCES
Issue: *Neuroimmunomodulation in Health and Disease*

TGF-β neutralization abrogates the inhibited DHEA production mediated by factors released from *M. tuberculosis*–stimulated PBMC

Luciano D'Attilio,[1] Verónica V. Bozza,[1] Natalia Santucci,[1] Bettina Bongiovanni,[1] Griselda Dídoli,[1] Stella Radcliffe,[2] Hugo Besedovsky,[3] Adriana del Rey,[3] Oscar Bottasso,[1] and María Luisa Bay[1]

[1]Institute of Immunology, School of Medical Sciences, National University of Rosario, Rosario, Argentina. [2]Central Laboratory Centenary Provincial Hospital, Rosario, Argentina. [3]Institut für Physiologie und Pathophysiologie, Marburg, Germany

Address for correspondence: Oscar Bottasso, Instituto de Inmunología, Facultad de Ciencias Médicas, Universidad Nacional de Rosario, Santa Fe 3100, 2000 Rosario, Argentina. oscarbottasso@yahoo.com.ar

Supernatants (SN) from cultures of peripheral blood mononuclear cells (PBMC) of tuberculosis (TB) patients inhibit dehydroepiandrosterone (DHEA) secretion by the adrenal cell line NCI-H295R. To analyze whether TGF-β is involved in this effect, SN of PBMC from healthy controls or patients with severe TB infections, stimulated or not with *Mycobacterium tuberculosis* (Mtb SN), were added to adrenal cells under basal conditions or following stimulation with forskolin. Cortisol and DHEA concentrations were evaluated in supernatants of the adrenal cells cultured with or without the addition of anti-TGF-β. Treatment with Mtb SN from TB inhibited DHEA production, and this effect was reversed when SN were treated with anti-TGF-β. The increase in cortisol production induced by SN from TB patients was not affected by TGF-β neutralization. Mediators released during the anti-TB immune response differentially modulate steroid production by adrenal cells, and TGF-β is a cytokine implicated in the inhibition of DHEA production observed in TB.

Keywords: dehydroepiandrosterone; cortisol; TGF-β; NCI-H295R cell line; tuberculosis

Introduction

Tuberculosis (TB) can be traced back for thousands of years and has probably threatened humanity since its origin. It is estimated that, at present, one-third of the world's population harbors the TB bacillus *Mycobacterium tuberculosis*, and that 10% of the infected individuals will develop active TB during their lifetime. This makes TB the single leading cause of death among all infectious diseases.[1] The infection affects mainly the lungs and begins as a nonspecific inflammatory reaction in the alveoli, followed by a gradual migration of macrophages, T cells, and fibroblasts that form a parenchymal granulomatous reaction directed at limiting mycobacterial spreading.[2] Primary infection is mostly self-resolving, although some bacilli persist in the tissues for months or decades, in a nonreplicative

state.[3] Most cases of secondary TB (5% of the infected people) are due to reactivation of old lesions, with pulmonary infiltrates and often parenchymal lung destruction, resulting in cavitary lesions. The affected areas include a few foci in the upper lobes and bilateral lung involvement, characterized by intense inflammation, tissue destruction, and fibrosis.[2,3] The type of active disease that develops is dictated generally by the state of the host's immune system. Infection in immunologically competent subjects induces early and late immune responses that ultimately destroy most tubercle bacilli and usually prevent development of clinical disease.[4,5] The first line of host defense against mycobacteria is conferred by macrophages through nonspecific mechanisms, such as phagocytosis, and production of different cytokines. If the infection cannot be controlled, T cells become involved two to three

doi: 10.1111/j.1749-6632.2012.06644.x
Ann. N.Y. Acad. Sci. 1262 (2012) 1–9 © 2012 New York Academy of Sciences.

weeks later. T cells contribute to defense against mycobacteria by producing cytokines, including interferon gamma (IFN-γ) and interleukin 2 (IL-2). Due to its substantial macrophage-activating effects, IFN-γ contributes to protective immunity against pathogens by activating macrophages to more effectively eliminate microorganisms, whereas IL-2 supports the proliferation of activated lymphocytes.[6] Although TNF-α is involved in *M. tuberculosis* clearance,[7] it may also account for several common features of TB, such as fever, wasting, and tissue damage.[8–10]

In agreement with the downregulatory role of anti-inflammatory cytokines in infections by intracellular pathogens,[2,6] increased IL-4 and IL-10 production has also been detected in patients with severe TB or anergic disease cases.[11–14] The proposal that cytokines such as IFN-γ play an essential role in the control of the *M. tuberculosis* infection is also reinforced by studies from our laboratory and those of others, indicating that patients with less severe forms of pulmonary TB have a predominant IFN-γ response, whereas the production of IL-4, IL-10, and TGF-β mediators, or potentially toxic compounds, predominate in aggravated disease.[15]

In addition to their immunological effects, many cytokines produced during the immune response also influence several neuroendocrine mechanisms, for example, activation of the hypothalamus–pituitary–adrenal (HPA) axis, as part of a well-regulated defense reaction.[16–18] Inflammatory cytokines can stimulate the synthesis of corticotrophin-releasing hormone in the hypothalamus, leading to the release of adrenocorticotropin hormone and the subsequent production of adrenal steroids, glucocorticoids (GCs), and dehydroepiandrosterone (DHEA),[16–18] or they can directly influence the activity of the adrenal gland.[19,20] At physiologic concentrations, GCs are able to shift the immune response from a proinflammatory to an anti-inflammatory cytokine pattern, and can also facilitate humoral immune responses, partly by inhibiting IFN-γ–producing cells.[16–18] Conversely, DHEA stimulates T helper functions by enhancing the capacity of activated T cells to produce IL-2, thus counteracting the inhibitory effect of GC on IL-2 synthesis. At the same time, DHEA seems to synergize with GC anti-inflammatory effects, as this adrenal androgen, like GC, is has potent antiphlogistic activity.[21]

Studies in *M. tuberculosis*–infected mice have shown that stimulation of the HPA axis contributes to disease reactivation.[22] Increased GC levels in *M. tuberculosis*–infected A/J mice were also found to counteract the development of protective Th1 responses,[23] whereas administration of DHEA and androstenediol, a steroid derivative, improves the course of experimental TB infection.[24] Work from the same group also documented important adrenal changes during experimental TB.[25] *In vitro* studies demonstrated that treatment of peripheral blood mononuclear cells (PBMC) from TB patients with physiological concentrations of cortisol inhibits mycobacterial antigen–driven lymphoproliferation and IFN-γ production, whereas DHEA suppresses TGF-β production.[26]

To better approach the clinical situation, we evaluated several immune and endocrine parameters in untreated TB patients with different degrees of lung involvement. Patients were HIV-negative, newly diagnosed, and presented with mild, moderate, or advanced disease. As our laboratory and those of others have shown,[14,27] IFN-γ, IL-10, IL-6, and growth hormone plasma levels were increased in TB patients, compared with healthy controls, in parallel with modest increases in concentrations of cortisol, estradiol, prolactin, and thyroid hormones and a profound decrease in testosterone and DHEA levels.[28] Because of the well-known effects of cytokines on endocrine functions,[16–18] the presence of a partly skewed profile of hormonal alterations may be related to cytokines released during the TB-specific immune response; consistent with this, supernatants of *M. tuberculosis*–stimulated PBMCs from TB patients significantly inhibited DHEA secretion by a human adrenal cell line.[28] TGF-β may be among the cytokines potentially involved in this effect because this cytokine can inhibit DHEA production,[29] and large amounts of TGF-β are detected in supernatants of cultures of *M. tuberculosis*–stimulated PBMCs from TB patients.[30] The study presented here provides evidence that treatment of these supernatants with anti-TGF-β–specific antibodies reverses the inhibition of DHEA production by adrenal cells.

Materials and methods

Study groups
Patients (one female and five males) with no HIV coinfection and newly diagnosed lung TB were

recruited for these studies. Diagnosis of *M. tuberculosis* infection was based on clinical and radiological data together with the identification of TB bacilli in sputum. The age of the patients ranged from 26 to 62 years (42.6 ± 20.1, mean ± SD, years). Disease severity was determined by radiological pattern and was classified as advanced TB. The control population was composed of seven healthy volunteers (healthy controls (HCo), three females, four males) of comparable age without any known prior contact with TB patients. None of the HCo had clinical or radiological evidence of active pulmonary TB, of any other respiratory disease, or of acute, chronic, or immunocompromising diseases or therapies. Additional exclusion criteria included diseases that affect the adrenal glands, the HPA or hypothalamus–pituitary–gonadal axes, corticosteroid treatment, pregnancy, and age below 18 years. Blood samples were obtained from all donors at entry into the study, and in the case of TB patients, before initiation of anti-TB treatment. This work was approved by the Ethical Committee of the Facultad de Ciencias Medicas, Universidad Nacional de Rosario. Participants were enrolled upon obtaining written consent.

Mononuclear cell isolation and in vitro stimulation

PBMC were isolated from freshly obtained EDTA-treated blood. In brief, blood was diluted 1:1 with culture medium (CM): RPMI 1640 (PAA Laboratories GmbH, Austria) containing standard concentrations of L-glutamine, penicillin, and streptomycin. The cell suspension was layered over a Ficoll-paque plus gradient (density 1.077, Amersham Biosciences, Piscataway, NJ) and centrifuged at 400 *g* for 30 min at room temperature (19–22 °C). PBMC recovered from the interface were washed three times with CM and resuspended in CM containing 10% of heat-inactivated pooled normal AB human serum (RPMI, PAA Laboratories GmbH). Cells were cultured in quadruplicate in flat-bottomed microtiter plates (5×10^6 cells/well in 1 mL) with or without the addition of whole sonicated, heat-killed H37Rv *M. tuberculosis* (Mtb; 8μg/mL, kindly provided by J.L. Stanford, London). PBMC cultures were incubated for 36 h at 37 °C in a 5% CO_2 humidified atmosphere.

Human adrenal cell line NCI-H295R cultures

The human adrenal cell line NCI-H295R (kindly provided by M. Ehrhart-Bornstein, Dresden, Germany)[31] was cultured in DMEM/F12 medium supplemented with L-glutamine and HEPES (Gibco), $NaHCO_3$ (1.2 g/L), insulin (379.47 ng/mL), hydrocortisone (3.625 ng/mL), estradiol (2.724 ng/mL), transferrin (10 μg/mL), selenite (5 ng/mL) (all from Sigma-Aldrich), penicillin (100 U/mL), streptomycin (100 μg/mL; Biochrom, Germany), and 2% heat-inactivated FCS (Gibco). Cells (70,000 cells/cm²) were cultured until 60–70% confluence (1.5×10^5 cells/well/400μL CM) in flat-bottom 24-well plates (Corning Costar, Cambridge, MA). Five days later, when cells were at the exponential growth phase, the medium was removed and 200 μL from pools of supernatants obtained from Mtb-stimulated or unstimulated PBMC and 200 μL of fresh medium were added to the cells. In a series of cultures, forskolin (FK; 2×10^{-5} M; Tocris, Biotrend, Germany) was used to stimulate the adrenals cells. Each treatment was assayed in quadruplicate. Supernatants from these cultures were obtained after 48 h and frozen at −20 °C until used for hormone determinations.

In summary, adrenal cells received either (1) CM alone (RPMI, as described in the previous paragraph); (2) supernatants from PBMC obtained without further stimulation (basal); or (3) supernatants from mycobacterial antigen-stimulated PBMC (Mtb SN).

TGF-β neutralization

In a first step, adrenal cell cultures at the exponential growth phase were treated with recombinant TGF-β (Sigma Aldrich, St. Louis, MO) at a range of concentrations corresponding to the amounts detected in TB patients and HCo (4,338 to 9,916 pg/mL). This set of experiments demonstrated that 6,100 pg/mL of TGF-β, the average concentration in supernatants of antigen-stimulated PBMC from patients with a severe disease, caused a 64% inhibition of DHEA production compared with untreated cultures (data not shown). Thus, this concentration was used as reference to determine the dilution of the anti-TGF-β antibody (Santa Cruz Biotechnology, CA) necessary to neutralize TGF-β in the supernatants. On the basis of the results of pilot neutralization studies, the relations used in the final experiments were 25:1, 10:1, and 1:1 (a 1:1 ratio corresponds to 6,100 pg/mL anti-TGF-β: 6,100 pg/mL TGF-β). Parallel cultures in which adrenal cells were treated only with the various anti-TGF-β concentrations were also performed for comparison purposes.

Hormone determinations

Cortisol and DHEA concentrations in culture supernatants of adrenal cells were determined using commercially available ELISA kits, according to the instructions of the manufacturer (DRG Systems, Marburg, Germany). The detection limits were 2.5 ng/mL for cortisol and 0.1 ng/mL for DHEA.

Statistical analysis

Data are shown as mean ± SEM of four independent determinations (corresponding to four individual cultures per group). Statistical comparisons were performed by the Kruskall–Wallis and Mann–Whitney U tests. $P < 0.05$ was considered statistically significant.

Results

TGF-β concentration was determined in supernatants of 36-h cultured PBMC from TB patients and controls. PBMC were either nonstimulated or stimulated with Mtb. Basal and *M. tuberculosis*–driven TGF-β production by PBMC from TB patients (5785.5 ± 797.6 pg/mL and 6757.7 ± 790.5 pg/mL, respectively) was higher than in cultures from HCo (4404.2 ± 1096.5 and 5692.1 ± 994.2 pg/mL, respectively), but the trend did not reach statistical significance.

After confirming that the human adrenal cell line NCI-H295R (with or without FK stimulation)

efficiently grows in the medium used to obtain the PBMC supernatants, pools of supernatants obtained from Mtb-stimulated or unstimulated PBMC of HCo and TB patients were added to the adrenal cells. Forty-eight hours later, supernatants from the human adrenal cell line were collected to determine the concentration of the adrenal steroids DHEA and cortisol. As seen in Figure 1, SN from unstimulated PBMC from HCo enhanced unstimulated and FK-stimulated DHEA secretion when compared with corresponding controls, but the secretion of the hormone decreased when adrenal cells were incubated with supernatants derived from Mtb-stimulated PBMC (Fig. 1A). Supernatants from Mtb-stimulated PBMC from TB patients also decreased DHEA production (Fig. 1B). However, and opposite to the effect of supernatants obtained from healthy donors, the supernatants from PBMC of TB patients that had not been further stimulated with the mycobacterial antigen *in vitro* exerted an inhibitory effect when adrenal cells were stimulated by FK.

Furthermore, it should be noted that SN from nonstimulated PBMC from HCo induced a fourfold increase in DHEA production by FK-stimulated adrenal cultures relative to the effect on spontaneous hormone release, whereas treatment with comparable SN from TB patients increased DHEA production by two-fold only.

Nonstimulated and FK-stimulated cortisol secretion by adrenal cells was augmented by SN from

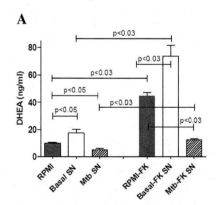
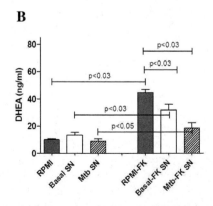

Figure 1. Effect of supernatants of PBMC on DHEA production by the human adrenal cell line NCI-H295R. Supernatants from cultures of PBMC stimulated with *M. tuberculosis* antigen (Mtb SN) or not (basal SN) from healthy controls (panel A) and TB patients (panel B) were collected. The supernatants were added to NCI-H295R adrenal cells. Another series of adrenal cell cultures received the supernatants together with forskolin (FK) to stimulate basal hormone production. The concentration of DHEA in the medium in which adrenal cells were cultured was determined 48 h later in duplicate. Bars and lines represent means ± SEM of four individual culture/group. Horizontal lines indicate comparisons between groups and statistically significant differences.

PBMC from both HCo and TB patients that were not further stimulated *in vitro* with the mycobacteria sonicate (Fig. 2). SN from Mtb-stimulated PBMC of HCo and TB patients induced a significant increase in cortisol secretion by adrenal cells that were not exposed to FK. Interestingly, the supernatants derived from Mtb-stimulated PBMC of TB patients were effective in significantly increasing cortisol secretion even when adrenal cells were already stimulated by FK (Fig. 2).

We then proceeded to analyze the production of DHEA and cortisol by adrenal cells when the PBMC supernatants had been treated with anti-TGF-β antibodies. Treatment of the adrenal cells with the mycobacterial antigen or anti-TGF-β antibodies alone did not affect DHEA production, independent of whether FK was also added. For a better appreciation of the effect caused by TGF-β neutralization, only results obtained in FK-stimulated adrenal cells are shown in Figure 3.

Treatment with anti-TGF-β reversed the inhibitory effect of SN from Mtb-stimulated PBMC from HCo and TB patients on DHEA production. In the case of SN from HCo, such an effect was already achieved when anti-TGF-β antibody was added at a 10:1 relation. However, the amount of antibody required to abrogate the inhibitory effect mediated by SN from TB patients had to be increased to 25:1. A trend to normalize DHEA production was also seen in adrenal cells exposed to SN from unstimulated PBMC of TB patients treated with anti-TGF-β (25:1 ratio).

When analyzing cortisol production, cultures exposed to SN from Mtb-stimulated PBMC from HCo treated with anti-TGF-β released significantly more cortisol than their untreated counterparts only when the antibody was used at a relation 10:1 (Fig. 4). No major differences in cortisol production were detected in cultures treated with SN from TB patients, independent of whether anti-TGF-β was added (Fig. 4).

Discussion

The host response to an infectious challenge involves a generalized defense reaction, characterized by changes in immune, metabolic, endocrine, and neural functions aimed at inhibiting pathogen growth and the accompanying inflammation.[32,33] From a teleological standpoint, these neuroimmunoendocrine changes may have an important adaptive value, but when the immune response fails to eradicate the pathogen, a chronic infection is established, leading to misdirected responses with harmful consequences.

Alterations in the concentration of adrenal steroids in patients with pulmonary TB, such as increased cortisol and decreased DHEA levels, are likely to influence the immune response, possibly by contributing to the gradual loss of the effect of Th1 cytokines.[15] In fact, increased levels of GCs

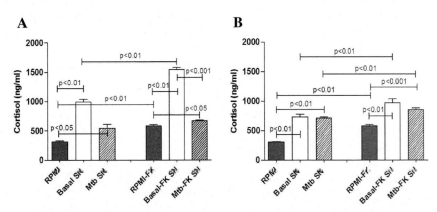

Figure 2. Effect of supernatants of PBMC on cortisol production by the human adrenal cell line NCI-H295R. Supernatants from cultures of PBMC stimulated with *M. tuberculosis* antigen (Mtb SN) or not (basal SN) from healthy controls (panel A) and TB patients (panel B) were collected. The supernatants were added to NCI-H295R adrenal cells. Another series of adrenal cell cultures received the supernatants together with forskolin (FK) to stimulate basal hormone production. The concentration of cortisol in the conditioned medium of the adrenal cells was determined 48 h later by duplicate. Bars and lines represent means ± SEM of four individual culture/group. Horizontal lines indicate comparisons between groups and statistically significant differences.

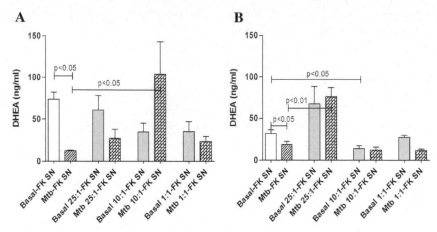

Figure 3. Effect of neutralizing TGF-β in supernatants of PBMC on DHEA production by the human adrenal cell line NCI-H295R. Supernatants from cultures of PBMC stimulated or not with *M. tuberculosis* antigen (Mtb SN and basal SN, respectively) from healthy controls (panel A) and TB patients (panel B) were collected. The supernatants were treated with anti-TGF-β at a ratio of 25:1, 10:1, or 1:1 (a ratio 1:1 corresponds to 6,100 pg/mL anti-TGF-β: 6,100 pg/mL TGF-β) and added to NCI-H295R adrenal cells cultured with forskolin (FK). The concentration of DHEA in the conditioned medium was determined in duplicate 48 h later. Bars and lines represent the mean ± SEM of four individual cultures/group. Horizontal lines indicate comparisons between groups and statistically significant differences.

exert anti-inflammatory effects, favoring an immune response that would be inefficient against intracellular pathogens.[34,35] DHEA counteracts the inhibitory effects of GCs, but also exerts potent anti-inflammatory actions.[21] Thus, the altered cortisol/DHEA relation, at the expense of the markedly reduced levels of DHEA, will favor an inhibition of cellular-mediated immune responses. The potential repercussion of the cortisol/DHEA balance on immune perturbations during TB was recently investigated by analyzing the relation between cortisol and DHEA levels and the *in vitro* immune response to mycobacterial antigens of PBMCs from patients with active TB. We have found that plasma DHEA levels are positively correlated with IFN-γ values, whereas an inverse correlation between the cortisol/DHEA ratio and IFN-γ levels was detected.[36]

The findings reported here extend previous observations that mediators released at different phases of the anti-TB immune response differentially modulate steroid production by adrenal cells.[28] We found that the effects vary depending on the source of the supernatants and the steroid hormone under analysis. Although SN of nonstimulated PBMC from healthy controls enhanced DHEA secretion, SN from Mtb-stimulated mononuclear cells exerted the opposite effect. Cells from TB patients also produced factors that inhibit DHEA

production by FK-stimulated adrenal cells and, interestingly, without further *in vitro* stimulation of the PBMC with mycobacterial antigens. Thus, mycobacterial stimulation induces the production of mediators capable of reducing DHEA secretion by adrenal cells. In the case of TB patients, *in vivo* infection with the bacilli is enough to stimulate PBMC to release such products *in vitro*. As opposed to the effects on DHEA production, PBMC supernatants from HCo and TB patients showed a general trend of increased cortisol production, though this increase was less evident in cultures treated with SN of nonstimulated PBMC from TB patients (Fig. 2).

Our studies also indicate that TGF-β is implied in the inhibition of DHEA production by adrenal cells. In fact, treatment with anti-TGF-β abrogated the inhibitory effects of SN from stimulated PBMC from HCo and TB patients on DHEA release. However, higher concentrations of the neutralizing antibody were necessary to reverse the inhibition caused by products from cells obtained from patients with active TB. These results agree with the finding that culture supernatants from PBMC of TB patients contained larger amounts of TGF-β than those from HCo counterparts. We cannot explain at present why a higher concentration of the antibody did not abrogate the inhibitory effect of supernatants from HCo. One possibility is that the relation of TGF-β

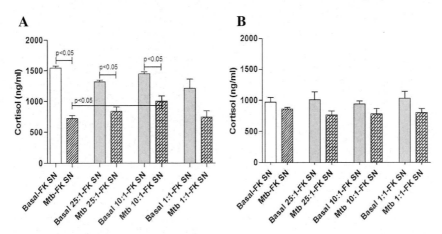

Figure 4. Effect of neutralizing TGF-β in supernatants of PBMC on cortisol production by the human adrenal cell line NCI-H295R. Supernatants from cultures of PBMC stimulated or not with *M. tuberculosis* antigen (Mtb SN and basal SN, respectively) from healthy controls (panel A) and TB patients (panel B) were collected. These supernatants were treated with anti-TGF-β at a ratio of 25:1, 10:1, or 1:1 (a ratio 1:1 corresponds to 6,100 pg/mL anti-TGF-β: 6,100 pg/mL TGF-β) and added to NCI-H295R adrenal cells cultured with forskolin (FK). The concentration of cortisol in the conditioned medium was determined in duplicate 48 h later. Bars and lines represent means ± SEM of four individual culture/group. Horizontal lines indicate comparisons between groups and statistically significant differences.

antibody to TGF-β might be critical, and that higher concentrations result in prozone-like effects.

TGF-β is a key cytokine in TB immunopathology because it inhibits macrophages and downregulates IFN-γ production.[37] Our previous studies showed that *M. tuberculosis*–stimulated PBMCs of TB patients with a severe form of the disease produce more TGF-β than those from healthy controls or from patients with a mild or moderate disease.[30] Extending these findings, we now add evidence involving TGF-β in the immunoendocrine communication during TB, particularly in the impaired DHEA secretion that these patients present.

A parallel, interesting finding was that neutralization of TGF-β in supernatants of Mtb-stimulated PBMC from HCo resulted in a relatively small but significant increased cortisol production by adrenal cells. These results indicate that Mtb-stimulation of PBMC from healthy donors triggers the production of soluble factors capable of increasing cortisol production, an activity that is manifested when TGF-β is neutralized. The fact that no such effect on cortisol production is observed when the supernatants are derived from TB patients might indicate that these supernatants are qualitatively different, and that other products, which are different from TGF-β, are responsible for the increase in cortisol production by adrenal cells. This observation adds an extra level of complexity to the intricate network

of immunoneuroendocrine interactions. Furthermore, the studies reported here show direct effects of immune products on adrenal cells. However, the actions of cytokines exerted upstream from this gland on the other components of the HPA axis have to be considered.

Conclusions

In chronic infectious diseases such as TB, excessive and/or prolonged cytokine production may affect immune–endocrine communication, favoring the establishment of an adverse state, characterized by important alterations in essential biological functions, together with perpetuated tissue damage. We have shown that mediators released during different phases of the anti-TB immune response differentially modulate steroid production by adrenal cells. Furthermore, to the well-known immunopathological role of TGF-β during TB, the results reported here suggest that this cytokine may be an important link in the communication between the immune and the endocrine systems, in particular, as a mediator of the disturbed cortisol/DHEA balance.

Acknowledgments

We thank J. Stanford and M. Ehrhart-Bornstein for kindly providing the *M. tuberculosis* antigen and the cell line NCI-H295R, respectively. This work was supported by Grants from the FONCYT (BID

1728/OC-AR 5-25462) (BID PID 160 PAE 37245), and the German Research Council (to A.d.R.).

Conflicts of interest

The authors declare no conflicts of interest.

References

1. World Health Organization. 2009. *Global Tuberculosis Control: Epidemiology, Strategy.* World Health Organization. Geneva, Switzerland. Financing: WHO report 2009.
2. Flynn, J.L. 2004. Immunology of tuberculosis and implications in vaccine development. *Tuberculosis* **84:** 93–101.
3. Small, P.M. & P.I. Fujiwara. 2001. Management of tuberculosis in the United States. *N. Engl. J. Med.* **345:** 189–200.
4. Baumann, S., A.N. Eddin & S.H.E. Kaufmann. 2006. Progress in tuberculosis vaccine development. *Curr. Opin. Immunol.* **18:** 438–448.
5. Ottenhoff, T.H., F.A. Verreck, M.A. Hoeve & E. van de Vosse. 2005. Control of human host immunity to mycobacteria. *Tuberculosis* **85:** 53–64.
6. North, R.J. & Y.J. Jung. 2004. Immunity to tuberculosis. *Annu. Rev. Immunol.* **22:** 599–623.
7. Dheda, K., H. Booth, J.F. Huggett, *et al.* 2005. Lung remodeling in pulmonary tuberculosis. *J. Infect. Dis.* **192:** 1201–1210.
8. Cadranel, J., C. Philippe, J. Perez, *et al.* 1990. In vitro production of tumour necrosis factor and prostaglandin E2 by peripheral blood mononuclear cells from tuberculosis patients. *Clin. Exp. Immunol.* **81:** 319–324.
9. Flynn, J.L. , M.M. Goldstein, J. Chan, *et al.* 1995. Tumor necrosis factor-alpha is required in the protective immune response against *Mycobacterium tuberculosis* in mice. *Immunity* **2:** 561–572.
10. Condos, R., W.N. Rom, Y.M. Liu & N.W. Schluger. 1998. Local immune responses correlate with presentation and outcome in tuberculosis. *Am. J. Respir. Crit. Care Med.* **157:** 729–735.
11. Rook, G.A., R. Hernandez-Pando, K. Dheda & S.G. Teng. 2004. IL-4 in tuberculosis: implications for vaccine design. *Trends Immunol.* **25:** 483–488.
12. Boussiotis, V.A., E.Y. Tsai, E.J. Yunis, *et al.* 2000. IL-10 producing T cells suppress immune responses in anergic tuberculosis patients. *J. Clin. Invest.* **105:** 1317–1325.
13. van Crevel, R., E. Karyadi, F. Preyers, *et al.* 2000. Increased production of interleukin 4 by CD4+ and CD8+ T cells from patients with tuberculosis is related to the presence of pulmonary cavities. *J. Infect. Dis.* **181:** 1194–1197.
14. Balikó, Z., L. Szereday & J. Szekeres-Bartho. 1998. Th2 biased immune response in cases with active *Mycobacterium tuberculosis* infection and tuberculin anergy. *FEMS Immunol. Med. Microbiol.* **22:** 199–204.
15. Bottasso, O., M.L. Bay, H. Besedovsky & A. del Rey. 2007. The immuno-endocrine component in the pathogenesis of tuberculosis. *Scand. J. Immunol.* **66:** 166–175.
16. Chrousos, G.P. 1995. The hypothalamic-pituitary-adrenal axis and immune-mediated inflammation. *N. Engl. J. Med.* **332:** 1351–1362.
17. Turnbull, A.V. & C.L. Rivier. 1999. Regulation of the hypothalamic-pituitary-adrenal axis by cytokines: actions and mechanisms of action. *Physiol. Rev.* **79:** 1–71.
18. Besedovsky, H. & A. del Rey. 1996. Immune-neuro-endocrine interactions: facts and hypothesis. *Endocr. Rev.* **17:** 64–101.
19. Ehrhart-Bornstein, M., J.P. Hinson, S.R. Bornstein, *et al.* 1998. Intraadrenal interactions in the regulation of adrenocortical steroidogenesis. *Endocr. Rev.* **19:** 101–143.
20. Bornstein, S.R. & M. Ehrhart-Bornstein. 2000. Basic and clinical aspects of intraadrenal regulation of steroidogenesis. *Z. Rheumatol.* **59**(Suppl. 2): 12–17.
21. Dillon, J. 2005. Dehydroepiandrosterone, dehydroepiandrosterone sulfate and related steroids: their role in inflammatory, allergic and immunological disorders. *Curr. Drug Targets Inflamm. Allergy* **4:** 377–385.
22. Howard, A.D. & B.S. Zwilling. 1999. Reactivation of tuberculosis is associated with a shift from type 1 to type 2 cytokines. *Clin. Exp. Immunol.* **115:** 428–434.
23. Actor, J.K., C.D. Leonard, V.E. Watson, *et al.* 2000. Cytokine mRNA expression and serum cortisol evaluation during murine lung inflammation induced by *Mycobacterium tuberculosis. Comb. Chem. High Throughput Screen* **3:** 343–351.
24. Hernandez-Pando, R., M.D.L. Streber, H. Orozco, *et al.* 1998. The effects of androstenediol and dehydroepiandrosterone on the course and cytokine profile of tuberculosis in Balb/c mice. *Immunology* **95:** 234–241.
25. Hernandez-Pando, R., H. Orozco, J. Honour, *et al.* 1995. Adrenal changes in murine pulmonary tuberculosis; a clue to pathogenesis? *FEMS Immunol. Med. Microbiol.* **12:** 63–72.
26. Mahuad, C., M.L. Bay, M.A. Farroni, *et al.* 2004. Cortisol and dehydroepiandrosterone affect the response of peripheral blood mononuclear cells to mycobacterial antigens during tuberculosis. *Scand. J. Immunol.* **60:** 639–646.
27. Verbon, A., N. Juffermans, S. Van Deventer, *et al.* 1999. Serum concentrations of cytokines in patients with active tuberculosis (TB) and after treatment. *Clin. Exp. Immunol.* **115:** 110–113.
28. del Rey, A., C.V. Mahuad, V. Bozza, *et al.* 2007. Endocrine and cytokine responses in humans with pulmonary tuberculosis. *Brain Behav. Immun.* **21:** 171–179.
29. Lebrethon, M.C., C. Jaillard, D. Naville, *et al.* 1994. Effects of transforming growth factor-beta 1 on human adrenocortical fasciculata-reticularis cell differentiated functions. *J. Clin. Endocrinol. Metab.* **79:** 1033–1039.
30. Dlugovitzky, D., M.L. Bay, L. Rateni, *et al.* 1999. *In vitro* synthesis of interferon-γ, interleukin-4, transforming growth factor-β, and interleukin 1-β by peripheral blood mononuclear cells from tuberculosis patients: relationship with the severity of pulmonary involvement. *Scand. J. Immunol.* **49:** 210–217.
31. Haidan, A., U. Hilbers, S.R. Bornstein & M. Ehrhart-Bornstein. 1998. Human adrenocortical NCI-H295 cells express VIP receptors. Steroidogenic effect of vasoactive intestinal peptide (VIP). *Peptides* **19:** 1511–1517.
32. Baumann, H. & J. Gauldie. 1994. The acute phase response. *Immunol. Today* **15:** 74–80.
33. Besedovsky, H. & A. del Rey. 1992. Immune- neuroendocrine circuits: integrative role of cytokines. *Front. Neuroendocrinol.* **13:** 61–94.

34. Ramirez, F., D.J. Fowell, M. Puklavec, *et al.* 1996. Glucocorticoids promote a Th2 cytokine response by CD4+ T cells in vitro. *J. Immunol.* **156:** 2406–2412.

35. Padgett, D.A., J.F. Sheridan & R. Loria. 1995. Steroid hormone regulation of a polyclonal Th2 immune response. *Ann. N.Y. Acad. Sci.* **774:** 323–325.

36. Bozza, V.V., L. D'Attilio, C.V. Mahuad, *et al.* 2007. Altered Cortisol/DHEA ratio in tuberculosis patients and its relationship with abnormalities in the mycobacterial-driven cytokine production by peripheral blood mononuclear cells. *Scand. J. Immunol.* **66:** 97–103.

37. Toossi, Z. & J.J. Ellner. 1998. The role of TGF-β in the pathogenesis of human tuberculosis. *Clin. Immunol. Immunopathol.* **87:** 107–114.

Ann. N.Y. Acad. Sci. ISSN 0077-8923

ANNALS OF THE NEW YORK ACADEMY OF SCIENCES

Issue: *Neuroimmunomodulation in Health and Disease*

Changes in the immune and endocrine responses of patients with pulmonary tuberculosis undergoing specific treatment

Bettina Bongiovanni,[1] Ariana Díaz,[1] Luciano D'Attilio,[1] Natalia Santucci,[1] Griselda Dídoli,[1] Susana Lioi,[2] Luis J. Nannini,[3] Walter Gardeñez,[4] Cristina Bogue,[5] Hugo Besedovsky,[6] Adriana del Rey,[6] Oscar Bottasso,[1] and María Luisa Bay[1]

[1]Institute of Immunology, School of Medical Sciences, National University of Rosario, Rosario, Santa Fe, Argentina. [2]Central Laboratory, Centenary Provincial Hospital, Rosario, Santa Fe, Argentina. [3]Pulmonary Section, Eva Perón School Hospital, Granadero Baigorria, Santa Fe, Argentina. [4]Pulmonary Section, Centenary Provincial Hospital, Rosario, Santa Fe, Argentina. [5]Pulmonary Section, I. Carrasco Hospital, Rosario, Santa Fe, Argentina. [6]Institut für Physiologie und Pathophysiologie, Marburg, Germany

Address for correspondence: María Luisa Bay, Instituto de Inmunología, Facultad de Ciencias Médicas, Universidad Nacional de Rosario. Santa Fe 3100, 2000 Rosario, Argentina. bay.mahuad@yahoo.com.ar

We evaluated immune and endocrine status following antituberculosis treatment in HIV-negative patients with newly diagnosed tuberculosis (TB). Treatment led to a decrease in IL-6, IL-1β, and C-reactive protein levels. Cortisol levels decreased throughout the anti-TB treatment, particularly after 4 months, but changes were less pronounced than those seen in proinflammatory mediators. Specific therapy resulted in increased dehydroepiandrosterone (DHEA) levels, which peaked after 4 months and started to decline after 6 months of treatment, reaching levels below those detected at inclusion. In contrast, in most patients, dehydroepiandrosterone sulfate (DHEAS) levels remained unchanged, although a trend toward increased concentrations was observed in a few cases 3 months after the treatment was finished. Specific therapy also resulted in more balanced cortisol/DHEA and cortisol/DHEAS ratios. Etiologic treatment involves favorable immune and endocrine changes, which may account for its beneficial effects.

Keywords: tuberculosis; etiological treatment; immune-endocrine changes; cytokines; adrenal steroids

Introduction

Tuberculosis (TB), one of the most important infectious diseases worldwide, is caused by *Mycobacterium tuberculosis*, a facultative intracellular bacterium that is capable of surviving and persisting within host mononuclear cells. The bacillus is primarily transmitted via the respiratory route because the infection occurs mostly in the lungs, although the microorganism can seed any organ via hematogenous spread. The bacillus may be eliminated by the host's responses, but, in the majority of the persons infected, the bacillus persists and a clinically latent infection is established. Among infected individuals, the lifetime risk of developing clinical TB is about 10%,[1] presumably due to the lack of a timely and appropriate protective response.

The clinical outcome ranges from mild forms of the disease with minimal symptoms and low bacterial burden, to sputum smear-positive pulmonary TB with severe symptoms, extensive disease, and a high bacillary load.

The expression of the disease, as well as the susceptibility to TB infection *per se*, is influenced by the host response, which is carried out by innate and acquired immune mechanisms.[2] The chronic nature of this mycobacterial disease implies a protracted response together with the accompanying inflammation, which may result in metabolic and neuroendocrine changes, ultimately affecting the mechanisms of host defense. To analyze the immune-endocrine alterations present in TB, we have carried out a series of studies in newly diagnosed, untreated patients with different degrees

doi: 10.1111/j.1749-6632.2012.06643.x

Table 1. Levels of cytokines, C-reactive protein (CRP), and adrenal steroids in TB patients at the time of diagnosis and healthy controls

Mediators	HCo ($n = 20$)	TB ($n = 12$)	P value
IL-1β (pg/mL)	0.07 (0.07–0.07)	0.47 (0.20–0.80)	$P < 0.01$
IL-6 (pg/mL)	1.75 (0.54–3.54)	17.20 (6.00–49.55)	$P < 0.01$
CRP (mg/L)	4.05 (3.27–8.65)	45.52 (10.62–78.47)	$P < 0.001$
Cortisol (ng/mL)	181.3 (129.0–227.1)	313.3 (196.7–335.4)	$P < 0.01$
DHEA (ng/mL)	8.975 (5.205–10.500)	4.470 (2.558–5.693)	$P < 0.01$
DHEAS (ng/mL)	1530 (1120–2303)	820 (580–1160)	$P < 0.01$
Cortisol/DHEA	31.49 (18.21–37.38)	65.75 (33.96–114.00)	$P < 0.01$
Cortisol/DHEAS	0.1021 (0.0872–0.1918)	0.3058 (0.1490–0.4926)	$P < 0.01$

NOTE: Data are expressed as median (range 25–75% percentile).

of lung compromise. Plasma levels of interferon gamma (IFN-γ), interleukin (IL)-10, and IL-6 were increased, whereas testosterone and dehydroepiandrosterone (DHEA) levels were profoundly decreased in TB patients, and these alterations were more evident in patients with an advanced disease. Growth hormone concentrations were markedly elevated in patients, in parallel to modest increases in the concentrations of cortisol, estradiol, prolactin, and thyroid hormones.[3] These endocrine changes may partly account for the deficient control of the inflammatory response and the gradual loss of protective responses that TB patients present with disease progression. In addition to its anti-inflammatory effects, increased levels of glucocorticoids (GCs) during infections result in ineffective immune responses that fail to clear intracellular pathogens.[4–6] In turn, DHEA counteracts the Th2-promoting effect of GCs and, at the same time, also exerts anti-inflammatory effects.[7] Supporting these counteracting effects of adrenal steroids on the immune response, we have also detected that plasma DHEA levels in TB patients were positively correlated with the amount of IFN-γ present in supernatants from cultures of peripheral blood mononuclear cells stimulated *in vitro* with mycobacterial antigens. Furthermore, there was an inverse correlation between the *in vitro* production of this cytokine and the cortisol/DHEA ratio in plasma.[8]

In a more recent study, discriminant analysis allowed us to identify DHEA as one of the most significant variables that separate TB patients from healthy individuals.[9] Taken together, our data suggest that decreased DHEA levels in TB patients may be unfavorable for the course of the disease, not only for

the inefficient anti-infective immune response but also for the clinical status of patients.[8,9] To begin to explore this possibility, we have analyzed circulating levels of adrenal steroids in TB patients throughout the course of the specific treatment, together with additional mediators with recognized proinflammatory activity like IL-1β, IL-6, and C-reactive protein (CRP).

Materials and methods

Subjects

Twelve patients (one female and eleven males), with no HIV coinfection, who were diagnosed with lung TB on the basis of clinical and radiological findings and identification of TB bacilli in sputum, were analyzed in this study. Patients with multidrug resistant TB were excluded. The mean age of TB patients was 37 ± 17 years (mean $\pm$ SD). Patients had moderate ($n = 7$) or advanced ($n = 5$) disease according to the radiological findings, as previously described.[10] Anti-TB therapy consisted of 6 months of rifampicin and isoniazid, initially supplemented by 2 months of pyrazinamide and ethambutol. Twenty age-matched healthy subjects (HCo, 42 ± 14 years, 6 females, and 14 males) living in the same area and without antecedents of contact with TB patients were included as controls. Exclusion criteria for all participants included pathologies affecting the hypothalamus–pituitary–thyroid or gonadal axis, or direct compromise of the adrenal gland, pregnancy, age under 18, as well as systemic or localized pathologies requiring treatment with corticosteroids or immunosuppressants. All subjects were BCG vaccinated and provided written consent to

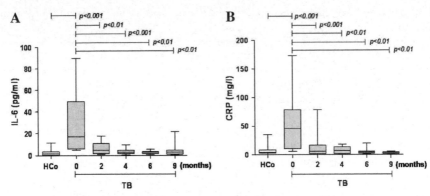

Figure 1. Levels of IL-6 and C-reactive protein (CRP) in plasma of TB patients during antituberculosis treatment and 3 months after its completion. Box plots show 25–75 percentiles of the results obtained in each group with maximum and minimum values. The line within the boxes represents the median values. HCo: healthy controls; $n = 20$; TB: patients with tuberculosis ($n = 12$). Time 0 indicates the results obtained at admission, before any treatment. Further samples were obtained 2, 4, and 6 months after starting the treatment, which was stopped at 6 months. An additional sample was obtained 3 months later (e.g., 9 months after entering in the study). Comparisons between groups (TB versus HCo) were performed by nonparametric methods (Kruskall–Wallis analysis of variance and Mann–Whitney U test). Paired comparisons at different times of antituberculosis treatment were done by Friedman analysis of variance and the Wilcoxon test. Horizontal lines above the boxes indicate the groups compared.

participate in these studies. The study was approved by the Ethical Committee of the Facultad de Ciencias Médicas, Universidad Nacional de Rosario and Centenario Hospital of Rosario.

Sample collection

Blood samples were obtained from TB patients at the time of diagnosis (before initiation of the treatment, T0) and 2, 4, and 6 months (T2, T4, T6) after starting the specific anti-TB treatment). An additional sample was obtained 3 months after the end treatment completion (T9). All samples were obtained between 8:00 and 9:00 a.m. using ethylene-diamine-tetraacetate (EDTA) as anticoagulant, and then centrifuged. Aprotinin (100 U/mL; Trasylol, Bayer, Germany) was added to the plasma shortly after collection, and the samples were preserved at –20 °C. One blood sample was obtained from each age- and sex-matched healthy control and processed in the same way.

Evaluation of immunological mediators

Levels of IL-1β, IL-6, and CRP were measured in plasma using commercially available high-sensitivity ELISA kits according to the instructions of the manufacturer. Detection limits were 0.06 pg/mL for IL-1β (Invitrogen Corporation, Camarillo, CA); 0.07 pg/mL for IL-6 (R&D Systems, Inc., Minneapolis, MN); and 2.5 mg/L for CRP (turbitest high-sensitivity, Wiener Lab, Rosario, Argentina)

Hormone determinations

Cortisol, DHEA, and DHEA-sulfate (DHEAS) were measured in plasma using commercially available ELISA kits according to the instructions of the manufacturer (DRG Instruments GmbH, Marburg, Germany). Detection limits were 2.5 ng/mL for cortisol; 0.1 ng/mL for DHEA; and 44 ng/mL for DHEAS.

Statistical analysis

Comparisons between groups (TB versus HCo) were performed by nonparametric methods (Kruskall–Wallis analysis of variance and the Mann–Whitney U test). Paired comparisons at different times of anti-TB treatment were done by Friedman analysis of variance and Wilcoxon test. Data were considered statistically significant when $P < 0.05$.

Results

At the time of diagnosis, TB patients displayed increased levels of IL-1β, IL-6, CRP, cortisol, cortisol/DHEA, and cortisol/DHEAS, and lower concentrations of DHEA and DHEAS when compared with HCo (Table 1). The same parameters were analyzed 2, 4, and 6 months after initiation of the treatment, which ended at this time, and 3 months later (i.e., 9 months after treatment initiation). As seen in Figure 1A and B, a sharp decrease in the levels of IL-6 and CRP was observed in TB patients after 2–4 months of specific treatment. This trend remained until the end of the study period. Similar results were

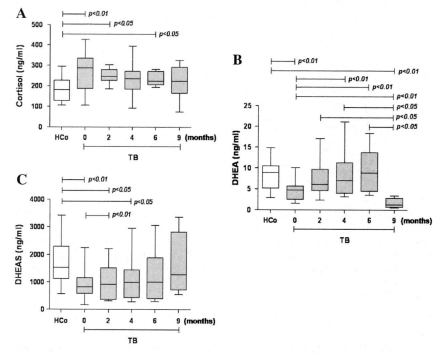

Figure 2. Levels of cortisol, DHEA, and DHEAS in plasma of TB patients during antituberculosis treatment and 3 months after its completion. Cortisol, DHEA, and DHEAS concentrations were determined in the same samples as described in Figure 1. Box plots show 25–75 percentiles of the results obtained in each group with maximum and minimum values. The line within represents the median values. Comparisons between groups (TB vs. HCo) were performed by nonparametric methods, (Kruskall–Wallis analysis of variance and the Mann–Whitney *U* test). Paired comparisons at different times of antituberculosis treatment were done by Friedman analysis of variance and the Wilcoxon test. Horizontal lines above the boxes indicate the groups compared.

obtained with IL-1β levels, which were significantly increased in newly diagnosed TB patients (median-rank, 0.47 pg/mL [0.20–0.80]) as compared with HCo (0.07 pg/mL [0.07–0.07], $P < 0.01$) and then decayed to values below the limit of detection.

In general terms, there was a decrease in cortisol plasma levels throughout the course of the anti-TB treatment, particularly at month 4 (Fig. 2A), but changes were less pronounced than those seen in the proinflammatory mediators. In contrast, patients undergoing anti-TB therapy had increased amounts of DHEA, reaching its highest levels at month 4, but values started to decline after 6 months of treatment lowering to levels even below the ones detected at time 0 (Fig. 2B). Regarding DHEAS, in most patients, its values remained within the same range along the treatment period, with a few cases showing a trend of higher concentrations at the nine-month evaluation (Fig. 2C). Further analysis on the balance between adrenal steroids revealed a significant decrease of cortisol/DHEA ratio nearly to normal values from month two and thereafter (Fig. 3A).

A similar but less-striking pattern was seen when analyzing the cortisol/DHEAS ratio (Fig. 3B).

We next calculated pairwise correlations between circulating mediators from TB patients. It was found that IL-6 positively correlated with cortisol/DHEAS at T9 ($r = 0.93$, $P < 0.01$). There was also a positive correlation between DHEA and DHEAS levels from T2 onward (T2: $r = 0.80$, $P < 0.01$; T4: $r = 0.64$, $P < 0.05$; T6: $r = 0.86$, $P < 0.01$; T9: $r = 1$, $P < 0.0001$). No bacilli were detected in the sputum of any of the patients one month after starting the treatment. The general physical condition and the body weight improved in all patients after 6 months of the standard therapy.

Discussion

Activation of the hypothalamus–pituitary–adrenal axis following stimulation of the immune response is usually initiated by cytokines with inflammatory activity, such as IL-1β and IL-6, leading to the ultimate increase of GCs and DHEA levels.[5,11] The main aim of this study was to explore the levels

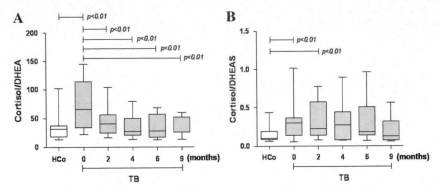

Figure 3. Cortisol/DHEA and cortisol/DHEAS ratios in plasma of TB patients during antituberculosis treatment and 3 months after its completion. The ratio between cortisol and DHEA, and cortisol and DHEAS, were calculated from the results shown in Figure 2. Box plots show 25–75 percentiles of the results obtained in each group with maximum and minimum values. The line within represents the median values. Comparisons between groups (TB versus HCo) were performed by nonparametric methods, (Kruskall–Wallis analysis of variance and the Mann–Whitney U test). Paired comparisons at different times of antituberculosis treatment were done by Friedman analysis of variance and the Wilcoxon test. Horizontal lines above the boxes indicate the groups compared.

of proinflammatory mediators, and of the steroids cortisol and DHEA during the course of the specific anti-TB treatment in patients. As shown in previous studies,[9,12] IL-6 and IL-1β levels were increased in TB patients at the time of diagnosis. Besides their proinflammatory effects, the concentration of both cytokines in plasma were negatively associated with body weight.[9] Studies in African TB patients also showed that increased IL-6 concentrations were correlated with loss of appetite.[13] Detrimental effects of IL-6 also extend to the immune response because IL-6 was found to inhibit macrophage responses to IFN-γ.[14]

In line with other reports,[15,16] the TB patients included in this study also displayed increased amounts of CRP at diagnosis, which highlights the important nonspecific systemic inflammation in this disease. In all patients, specific therapy was followed by a remarkable decrease in the levels of this compound, which fits well with the improved clinical status of the patients.

Confirming and extending former observations, active TB was characterized by low levels of DHEA and DHEAS in parallel to slightly increased levels of cortisol.[3,9] These alterations were also observed in several forms of physical stress, including trauma and chronic disabling inflammatory diseases.[17] This probably reflects a redistribution of the steroid flux to maintain cortisol production, despite a disturbed adrenal function, as an attempt to counteract the inflammatory response accompanying active dis-

ease. The increase in serum DHEA and DHEAS, the lowered cortisol/DHEA ratio, together with slightly changed cortisol levels during treatment, may reflect an adrenal improvement because of the disappearance of potentially toxic endogenous compounds synthesized in response to infection and a less inflammatory environment. Studies *in vitro* have documented that proinflammatory cytokines like TNF-α influence human adrenal steroidogenesis.[18]

Whatever the case, recovered levels of DHEA at the end of treatment may also contribute to the improved clinical condition and immune status given its potent anti-inflammatory effects and Th1-facilitating functions.[7]

In discussing results, it is interesting to note the reduced levels of DHEA exhibited by TB patients at 9 months in the presence of preserved amounts of DHEAS. Reasons for such findings may only be speculative. DHEA and its sulphate ester DHEAS are adrenal-derived steroids that can be interconverted. DHEAS converts to DHEA by the activity of DHEAS sulphatase, whereas the conversion of DHEA to DHEAS is mediated by a DHEA sulphotransferase. DHEAS is part of the hormone pool, has no effects, and constitutes a stable marker for DHEA availability. Following conversion to the biologically active DHEA in peripheral tissues, the hormone is intracellularly processed to yield active metabolites.[19] Hence, it may be that inhibition in DHEAS sulfatase activity reduced the transformation of the sulfatated mediator into DHEA, resulting in the reduced levels of DHEA that we detected. Such inhibition does not seem to be present during treatment with the anti-TB drugs rifampicin and isoniazid. Rifampicin is likely to affect the metabolism of adrenal steroids in the liver because of its effects on the hepatic CYP3A4 enzyme complex regulating key points in human steroidogenesis.[20] Studies on the adrenocortical function in active patients receiving either a rifampicin-based or a ciprofloxacin-based anti-TB regimen revealed no compromise of adrenal function.[21] Because evaluations in this study were performed the first 5 days of treatment, the potential effects of a prolonged intake need to be evaluated. It also remains to be established whether the reduced conversion of DHEAS to DHEA constitutes a particular feature of TB patients. Beyond the proper mechanisms, the present results suggest that specific therapy also involves a favorable qualitative change in the immune-endocrine profile accounting for the benefits of etiological treatment of tuberculosis.

Acknowledgments

This work was supported by grants from FONCYT (PID 160 PAE 37245) and the Ministry of Health and Secretariat for Science, Technology, and Innovation of the Santa Fe Province. We thank Wiener Labs for providing the CRP turbitest high-sensitivity kit and Ms. Daniela Flores for her technical assistance to patients.

Conflicts of interest

The authors declare no conflicts of interest.

References

1. Comstock, G.W. 1982. Epidemiology of tuberculosis. *Am. Rev. Respir. Dis.* **125:** 8–15.
2. Flynn, J.L. 2004. Immunology of tuberculosis and implications in vaccine development. *Tuberculosis* **84:** 93–101.
3. del Rey, A., C.V. Mahuad, V.V. Bozza, *et al.* 2007. Endocrine and cytokine responses in humans with pulmonary tuberculosis. *Brain Behav. Immun.* **21:** 171–179.
4. Van den Berghe, G. 2003. Endocrine evaluation of patients with critical illness. *Endocrinol. Metab. Clin. North Am.* **32:** 385 410.
5. Besedovsky, H.O. & A. del Rey. 1996. Immune-neuroendocrine interactions: facts and hypothesis. *Endocr. Rev.* **17:** 64–95.
6. Elenkov, I.J. & G.P. Chrousos. 1999. Stress hormones, Th1/Th2 patterns, pro/anti-inflammatory cytokines and susceptibility to disease. *Trends Endocrinol. Metab.* **10:** 359–368.

7. Dillon, J. 2005. Dehydroepiandrosterone, dehydroepiandrosterone sulfate and related steroids: their role in inflammatory, allergic and immunological disorders. *Curr. Drug Targets Inflamm. Allergy* **4:** 377–385.
8. Bozza, V.V., L. D'Attilio, C.V. Mahuad, *et al.* 2007. Altered Cortisol/DHEA ratio in tuberculosis patients and its relationship with abnormalities in the mycobacterial-driven cytokine production by peripheral blood mononuclear cells. *Scand. J. Immunol.* **66:** 97–103.
9. Santucci, N., L. DAttilio, L. Kovalevski, *et al.* 2011. A multifaceted analysis of immune-endocrine-metabolic alterations in patients with pulmonary tuberculosis. *PLoS One* **6:** e26363.
10. Mahuad, C., M.L. Bay, M.A. Farroni, *et al.* 2004. Cortisol and dehydroepiandrosterone affect the response of peripheral blood mononuclear cells to mycobacterial antigens during tuberculosis. *Scand. J. Immunol.* **60:** 639–646.
11. Turnbull, A.V. & C. Rivier. 1995. Regulation of the HPA axis by cytokines. *Brain Behav. Immun.* **9:** 253–275.
12. Mahuad, C., V. Bozza, S.M. Pezzotto, *et al.* 2007. Impaired immune responses in tuberculosis patients are related to weight loss that coexists with an immunoendocrine imbalance. *Neuroimmunomodulation* **14:** 193–199.
13. van Lettow, M., J.W.M. van der Meer, C.E. West, *et al.* 2005. Interleukin-6 and human immunodeficiency virus load, but not plasma leptin concentration, predict anorexia and wasting in adults with pulmonary tuberculosis in Malawi. *J. Clin. Endocrinol. Metab.* **90:** 4771–4776.
14. Nagabhushanam, V., A. Solache, L.M. Ting, *et al.* 2003. Innate inhibition of adaptive immunity: *Mycobacterium tuberculosis*-induced IL-6 inhibits macrophage responses to IFN-γ. *J. Immunol.* **171:** 4750–4757.
15. Choi, C.M., C.I. Kang, W.K. Jeung, *et al.* 2007. Role of the C-reactive protein for the diagnosis of TB among military personnel in South Korea. *Int. J. Tuberc. Lung Dis.* **11:** 233–236.
16. Breen, R.A., O. Leonard, F.M Perrin, *et al.* 2008. How good are systemic symptoms and blood inflammatory markers at detecting individuals with tuberculosis? *Int. J. Tuberc. Lung Dis.* **12:** 44–49.
17. Straub, R.H. & H.O. Besedovsky. 2003. Integrated evolutionary, immunological, and neuroendocrine framework for the pathogenesis of chronic disabling inflammatory diseases. *FASEB J.* **17:** 2176–2183.
18. Jäättelä, M., O. Carpen, U.H Stenman & E. Saksela. 1990. Regulation of ACTH-induced steroidogenesis in human fetal adrenals by rTNF-alpha. *Mol. Cell Endocrinol.* **68:** R31–R36.
19. Rosenfeld, R.S., L. Hellman & T.F. Gallagher. 1972. Metabolism and interconversion of dehydroisoandrosterone and dehydroisoandrosterone sulfate. *J. Clin. Endocrinol. Metab.* **35:** 87–193.
20. Katzung, B., S. Masters & A. Trevor. 2009. *Basic and Clinical Pharmacology*, 11th Ed. McGraw-Hill. New York.
21. Francois Venter, W.D., V.R. Panz, C. Feldman & B.I. Joffe. 2006. Adrenocortical function in hospitalised patients with active pulmonary tuberculosis receiving a rifampicin-based regimen—a pilot study. *S. Afr. Med. J.* **96:** 62–66.

Ann. N.Y. Acad. Sci. ISSN 0077-8923

ANNALS OF THE NEW YORK ACADEMY OF SCIENCES
Issue: *Neuroimmunomodulation in Health and Disease*

Sex steroids, immune system, and parasitic infections: facts and hypotheses

Karen Nava-Castro,[1] Romel Hernández-Bello,[2] Saé Muñiz-Hernández,[3] Ignacio Camacho-Arroyo,[1] and Jorge Morales-Montor[2]

[1]Departamento de Biología, Facultad de Química, Universidad Nacional Autonoma de Mexico. [2]Departamento de Inmunología, Instituto de Investigaciones Biomédicas, Universidad Nacional Autonoma de Mexico. [3]Instituto Nacional de Cancerología, México Distrito Federal, México

Address for correspondence: Jorge Morales-Montor, Departamento de Inmunología, Instituto de Investigaciones Biomédicas, Universidad Nacional Autónoma de México, AP, 70228 México D.F., México. jmontor66@biomedicas.unam.mx

It has been widely reported that the incidence and the severity of natural parasitic infections are different between males and females of several species, including humans. This sexual dimorphism involves a distinct exposure of males and females to various parasite infective stages, differential effects of sex steroids on immune cells, and direct effects of these steroids on parasites, among others. Typically, for a large number of parasitic diseases, the prevalence and intensity is higher in males than females; however, in several parasitic infections, males are more resistant than females. In the present work, we review the effects of sex hormones on immunity to protozoa and helminth parasites, which are the causal agents of several diseases in humans, and discuss the most recent research related to the role of sex steroids in the complex host–parasite relationship.

Keywords: estrogens; progesterone; androgens; protozoan; cestodes; immunity

Introduction

It is well known that sex steroids regulate a variety of functions such as growth, reproduction, and differentiation. More recently, the ability of sex steroids to regulate the immunological response against pathogenic agents has gained attention. The participation of sex steroids is evident in various parasitic diseases, including malaria, toxoplasmosis, cysticercosis, trypanosomiasis, and leishmaniasis, in which a steroid hormone regulation of the immune response has been described (see Table 1).[1–6] Parasites have developed diverse mechanisms of survival within the host, which facilitate the establishment of infection. These can be grouped into two types: those in which the immune response is evaded by strategies such as antigenic variation and molecular mimicry, and those in which the parasite exploits some system of the host to its benefit, producing its establishment, growth, or reproduction. Thus, *Naegleria fowleri* is capable of internalizing antigen antibody complexes from its surface with the dual benefit of gaining amino acids for their own metabolism and preventing the actions of the antibody.[7] Other pathogens, including *Chlamydia trachomatis* and *Coxiella burnetii*, have developed molecules that directly interfere with antigen processing and presentation.[8] A striking example of exploitation of host molecules is the ability of a number of parasites to use host-synthesized cytokines as indirect growth factors.

The ability of a parasite to differentially affect a female or a male of the same species (sexual dimorphism of an infection) can be due to the regulation of the immune response by sex hormones (Fig. 1). The comparatively sophisticated immune systems of vertebrates add complexity to host–parasite (H–P) interactions. Mammals sense and react with their innate and acquired immunological systems to the presence of a parasite, and the parasite is also sensitive and reactive to the host's immune system's effectors. Thus, it can be inferred that, in addition to their effects on sexual differentiation and reproduction, sex hormones

doi: 10.1111/j.1749-6632.2012.06632.x

Table 1. Sex-associated susceptibility and different sex steroid effects on several parasite infections

Parasite	Sex-associated susceptibility	Experimental condition	Effect	Refs.
Protozoa				
Toxoplasma gondii	F > M	Treatment with P4	Proliferation is not affected in macrophages	20
		Treatment with T4	Reduces parasites number and pathology	3, 21
		Treatment with 17β-estradiol, diethylstilbestrol, α-dienestrol	Increases development of brain cysts	22
		Treatment with 5α-dihydrotestoterone	No changes in development of brain cysts	22
		Ovarectomy	Reduces the development of tissue cysts	22
Plasmodium sp.	M > Fp > F	Treatment with P4	Increases susceptibility at infection	25
		Treatment with T4	Increases mortality	23–26
		Castration	Reduces the parasitemia and mortality	26
		Restitution with T4 after castration	Increases the mortality	23–24
		Gonadectomy male and female	Modulates immune response	22
Leishmania sp.	M > F	Treatment with E2	Induces higher resistance to infection in both male and female derived macrophages	30
		Treatment with T4	Increases infection in bone marrow–derived macrophages	32
		T4 administered to females	Increases susceptibility to infection	28, 29
		Castration male	Reduces susceptibility to infection	28
Trypanosoma cruzi	M > F	Treatment with E2	Decreases the number of trypomastogotes in blood	31
		Treatment with P4	Decreases the number of trypomastogotes in blood	31
		Treatment with 17 β-estradiol	Increases the mortality and parasitemia	34

Continued

Table 1. *Continued*

Parasite	Sex-associated susceptibility	Experimental condition	Effect	Refs.
		Gonadectomy	Eliminates sexual dimorphism	5, 30
		T4 replacement in GX males	Restores the parasitemia	5
		Castration of female	Reduces the number of blood parasites	33
		Orchidectomy	Induces lower parasitemia levels and higher resistance	1
Entamoeba histolytica	M > F ALA	Treatment with E2	Confers protection to infection	37, 39
			Trophozoite's proliferation *in vitro* is not modified	
	F > M intestinal infections	Treatment with T4	Favors migration from intestine to liver	38, 39
			Trophozoite's proliferation *in vitro* is not modified	
		Treatment with P4	Trophozoite's proliferation *in vitro* is not modified	39
		Gonadectomy	Reduces the development of ALA	
Helminthes				
Taenia crassiceps	F > M	Feminization process	Eliminates sexual dimorphism	44
		Treatment with E2	Decreases parasite growth	46
		Gonadectomy	Eliminates sexual dimorphism	47
Tenia solium	F > M	Treatment with P4	Reduces the number of adult worms	52
		Castration	Natural infection induces major prevalence of cysticercosis	53, 54

NOTE: F, females; Fp, pregnant females; M, males.

may also determine the differences between the genders regarding their immune response to the same antigenic stimulus. These differences include sexual dimorphism of the immune response and dimorphism associated with infection parameters.[9–11] The effects of sex hormones on functions of virtually all immune-cell types have been recently described,[12] including thymocyte maturation and selection, cellular transit, lymphocyte proliferation, expression of class II major histocompatibility complex molecules and receptors, and cytokine production.[13,14] Furthermore, the presence

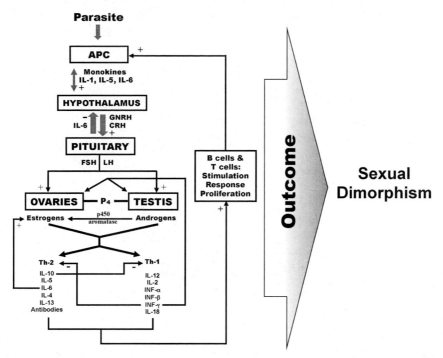

Figure 1. Immunoneuroendocrine network circuitry during parasitic diseases. The magnitude and complexity of the immunoendocrine network includes practically all sex hormones and many of the immunological components described. Some of the events in the immunoendocrinological network involve cellular reproduction and *de novo* synthesis of receptors. A node for strong hormonal modulation of acquired immunity in the immunoendocrine network is the proposed conflictive effect of androgens and estrogens acting upon the mutually controlled Th1/Th2 balance. The magnitude and complexity of the network includes practically all of the host's sexual hormones and many of the immunological components described, connecting among themselves and with the parasite's systems. Arrows (←, ↑, →, ↓) denote connections between nodes; each points to the direction of the signal. (−,+) signs refer to inhibiting or stimulatory effects. The final outcome of these interactions results in sexual dimorphism to parasite infection.

of sex steroid receptors on immune cells[12] indicates that one mechanism involved in sex steroid effects is through the interaction with either membrane or intracellular receptors (Fig. 2). Although sex hormones also exert their effects through nongenomic mechanisms by acting on cell surface receptors and triggering signaling cascades, it is currently accepted that the main route of biological activity occurs by the activation of specific intracellular receptors that function as ligand-activated transcription factors, and coordinate, after binding to their ligand, the expression of target genes (Fig. 2). The main mediators of sex steroid effects are estrogen receptors (ERs), which include two subtypes, ER-α and ER-β, each coded by different genes and whose predominating ligand is 17β-estradiol; progesterone receptors (PRs), which have two isoforms, PR-A and PR-B, generated from the same gene and whose main ligand is progesterone; and androgen receptors (ARs), coded by a single

gene and whose ligands are testosterone and dihydrotestosterone (DHT; Fig. 2). In addition, membrane receptors to sex steroids exist on immune cells. In fact, the binding between estradiol (E2) and its membrane ERs activates metabotropic glutamate receptor groups I and II.[15] ERs are able to bind to Src kinases through their highly conserved SH2 domains, which could modify the effect of ERK 1/2 on the phosphorylation pattern of several transcription factors.[16] Recently, three membrane PRs (mPRs), mPR-α, mPR-β, and mPR-γ, have been described and detected on T lymphocytes.[17] The mechanism of action of these mPRs is suggested to be through G_i-protein activation.[18] Previous findings have also revealed unconventional nongenomic surface receptors for testosterone in rat T cells.[19] Preliminary evidence indicates that in murine T cells, testosterone also induces a rapid rise in $[Ca^{2+}]^i$, presumably due to Ca^{2+} influx triggered by the binding of testosterone to receptors on the outer surface of

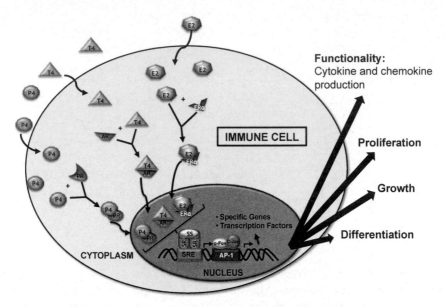

Figure 2. Genomic mechanisms; nuclear sex steroid receptors in immune cells. The figure indicates that one mechanism involved in sex steroid effects is through the interactions with their specific membrane or nuclear receptors in immune cells. The cells in which nuclear hormone sex steroid receptors have been described are T lymphocytes, B lymphocytes, macrophages, NK cells, CD8 cells, dendritic cells, mast cells, T_{reg} cells, and Th17 cells. The main route of biological activity occurs by the activation of specific intracellular receptors that function as ligand-activated transcription factors, and coordinate, after binding to their ligand, the expression of target genes. The main sex steroid hormone receptors (SSH) are estrogen receptors (ERs), which include two subtypes, ER-α and ER-β (main ligand is 17β-estradiol); progesterone receptors (PRs), which have two isoforms (PR-A and PR-B) generated from the same gene, and whose main ligand is progesterone; and androgen receptors (ARs), coded by a single gene with the main ligands being testosterone and dihydrotestosterone (DHT).

T cells (Fig. 2).[19] We will summarize some of the most-studied parasites in which sexual dimorphism has been found.

Protozoan parasites

Toxoplasma gondii

Infection with *T. gondii* is mainly acquired through the intake of cyst tissues (present in the meat of infected animals) or oocysts (released in the feces of infected cats). There is abundant evidence that sex steroid hormones affect the course of toxoplasmosis in humans and mice. In the case of progesterone, a direct effect of this hormone has been shown on the intracellular proliferation of the parasite and not through the macrophages.[20] Studies have investigated the effect that sex and sex hormones have on the immunity of the small intestine of mice during infection with tissue cysts. For example, infected female animals die significantly earlier than males, an effect associated with greater numbers of tachyzoites and necrosis in female animal small intestines; testosterone treatment of female mice reduced intestinal parasite numbers and pathology.[3] Kanková

et al. studied changes in the testosterone levels in the latent phase of toxoplasmosis in laboratory mice.[21] Using a relatively virulent strain, T38, the authors observed a decrease in testosterone levels in both male and female mice compared with uninfected animals, indicating that the infection induced a change in testosterone metabolism.[21] An increase in the number of brain cysts in mice can be induced by pharmacological concentrations of potent estrogenic compounds, including 17β-estradiol, diethylstilbestrol, and α-dienestrol. In contrast, androgenic compounds such as 5α-dihydrotestoterone produced no change in brain cyst formation.[22] In addition, ovariectomy of mice reduced the development of tissue cysts, whereas the administration of E2 exacerbated it.[22]

Plasmodium sp.

Several vertebrate hosts infected with *Plasmodium sp.* show pronounced sexual dimorphism,[23] which translates into higher parasitemia and mortality in males than in females.[24,25] Moreover, castration of male mice reduces parasitemia and mortality caused by *Plasmodium sp.*[26] This dimorphism is evident in humans, as men have higher parasitemia than women. It has also been observed that *P. falciparum* density increases during puberty in men but not in women, suggesting that circulating sex steroids may influence the parasite. Indeed, studies in rodents have confirmed that males are more likely to die after blood-stage malaria infection than are females. In addition, male mice infected with *P. chabaudi* are more susceptible to death than are females. Castration of male mice reduces mortality following infection with *P. chabaudi* or *P. bergei*, whereas exogenous administration of testosterone increases it.[23,25] The immunomodulatory effects of testosterone may underlie the increased susceptibility to *Plasmodium sp.* infection by males compared with females.[25] However, pregnant women are more susceptible than nonpregnant women to *P. falciparum* infections and frequently have a higher parasitemia.[24] This fact can be partially explained by the immunomodulatory properties of progesterone, which is increased in pregnant females.

Leishmania sp.

Leishmaniases show a wide range of clinical manifestations in humans, depending on the parasite species. The gender and age of the host are two important variables that affect the course of infection with *Leishmania sp.* Both male and female hormones can mediate sex-determined resistance and susceptibility to infection. In areas with endemic leishmaniasis, clinical disease is more frequent among males than females.[27] Experimental infections using parasites of the subgenus *Viannia* (which is the most common cause of cutaneous leishmaniasis in North America and South America) showed that infection of male hamsters resulted in significantly greater lesion size and severity than infection in female animals. The increased severity of pathology in male compared to female animals was associated with a significantly greater expression of proinflammatory cytokines at the site of the lesion.[28] It has also been reported that castration of males reduces the susceptibility to *L. major*, whereas the administration of testosterone to females increases it.[27] In endemic areas of leishmaniases produced

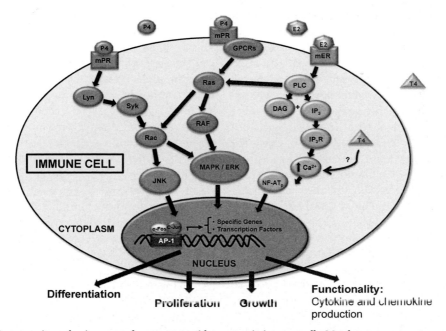

Figure 3. Nongenomic mechanisms; membrane sex steroid receptors in immune cells. Membrane receptors to sex steroids exist on immune cells. In B and T cells, the binding between estradiol (E2) and its membrane ERs activates metabotropic glutamate receptor groups I and II. ERs are able to bind to Src kinases through their highly conserved SH2 domains, which could modify the effect of ERK 1/2 on the phosphorylation pattern of several transcription factors. Three membrane progesterone receptors (mPRs), mPR-α, mPR-β, and mPR-γ, have been described and detected on T lymphocytes. The mechanism of action of these mPRs is suggested to be through G_i-protein activation. Nongenomic surface receptors for androgens (mAR) in rat T cells have also been described. Preliminary evidence indicates that in murine T cells, androgens induce a rapid rise in $[Ca^{2+}]^i$, presumably due to Ca^{2+} influx triggered by the binding of AR on the outer surface of T cells.

by *L. mexicana*, men have more frequent infections than women. Also, a sex-associated difference in susceptibility to *L. mexicana* has been found in DBA/2 mice. This difference was reported to be associated to the immune response: resistance in females is due to an induction of a higher production of IFN-γ, whereas susceptibility in males is associated with TNF-α.[5] Interestingly, the effect of E2 was evaluated *in vitro* on macrophages derived from male and female DBA/2 mice; under normal conditions, male-derived macrophages showed higher infection rates and contained more parasites compared with female-derived macrophages.[29] When macrophages were treated with E2, macrophages derived from male and female animals were more resistant to infection, and chronic treatment with E2 induced leishmanicidal activity in macrophages.[29] On the other hand, testosterone treatment increased the *L. donovani* infection rate in mouse bone marrow–derived macrophages,[30] probably through a nongenomic pathway.[31]

Trypanosoma cruzi

In the case of *T. cruzi*, several animal models have shown higher levels of parasitemia, more extensive tissue invasion, greater weight loss, and shorter survival in males than in females.[5,29] Such differences between sexes disappeared in gonadectomized (GX) female mice, which resulted in increased *T. cruzi* parasitemia.[5,29] In addition, treatment of mice with E2 and progesterone decreased the number of blood trypomastigotes to the levels observed in control animals.[30] These results demonstrated that female gonadal hormones, especially estrogens, play a fundamental role in the resistance to *T. cruzi*. In contrast, GX males showed lower parasitemia compared with control animals, and hormonal replacement with testosterone restored parasitemia to control levels, confirming the influence of male hormones in susceptibility to infection.[5] Castration of female mice produced significantly fewer blood parasites, higher lytic antibody percentage, higher splenocyte proliferation, and NO concentration compared with intact infected animals.[32] These findings indicate that steroid ablation influences the immune response to *T. cruzi* infection.[32] The treatment of infected C57/BL6 female mice at doses of 50–500 μg E2/mouse increased mortality and parasitemia; however, minor doses had no effect or even reduced both parameters. In the same study, the authors found that infection of mice during the metestrus phase of the estrous cycle, when E2 levels are low, presented with longer survival than mice infected during other phases of the cycle.[33] Orchiectomized mice, *Calomys callosus*, infected with the Y strain of *T. cruzi* had lower parasitemia levels than those obtained from intact males. Hormonal replacement, using testosterone in orchiectomized animals, induced similar values than those of the control group.[1] The trypomastigote lysis percentage varied through the course of infection according to hormonal status and number of parasites during the acute phase. Higher resistance with lower lysis indexes were observed after orchiectomy, compared with control males.[1] Even though there is no clear evidence of sexual dimorphism in humans, indirect observations suggest a lower severity and frequency of illness in women from *T. cruzi* infection.[31,32]

Entamoeba histolytica

Although information is scarce, epidemiological data have pointed out that sex hormones could play a role in the development of *E. histolytica* human infections, mainly acute liver abscess (ALA). It is well documented that in adults ALA is five- to sevenfold more prevalent in men than in women.[34,35] This difference is less clear in the case of intestinal amoebiasis; however, symptomatic intestinal amoebiasis in adults appears, in contrast to ALA, to be more common in women than in men.[33] These variations in susceptibility associated with sex suggest that sex hormones are involved in the development of amoebic infection. An early report proposed that estrogens confer protection against development of ALA in hamsters.[36] The only reported effect for progesterone was the increase in positive migration of trophozoites from the intestine to the liver in gerbils.[37] More recently, differences within sex in the control of ALA were reported in C57BL/6 mice. In contrast to males, females rapidly cleared the parasites, recruited higher numbers of natural killer T cells to the infection site, and produced higher levels of IFN-γ, suggesting a sexual dimorphism.[35] Moreover, we have reported that trophozoite *in vitro* treatment with sex hormones, such as progesterone, E2, and testosterone, did not affect trophozoite's proliferation.[38] According to the experimental evidence presented previously that supports that host sex is relevant during amoebiasis, it has been hypothesized that differences between sexes in hamsters could not really be detected due to the overactivation of the Th1 immune response, which masks differences associated with sex hormones. A recent study showed a differential effect of gonadectomy on hamster development of ALA that was absent in 50% of male and 15% of female GX hamsters, compared with 100% infection in non-GX controls. The protection against ALA in GX hamsters was concomitant with a comparatively scarce inflammatory infiltrate and necrosis surrounding clusters of trophozoites in the liver tissue, as well as a lack of response of spleen cells to concanavalin A evaluated in proliferation assays. Immunohistochemistry of liver sections showed a strong Th1 responses in non-GX animals, although GX females and males exhibited a Th2 and Th3 profile of cytokines, respectively, suggesting that protection against ALA following GX should be related to a downregulation of liver Th1 response during amoebic infection.[39]

Taenia crassiceps

Murine intraperitoneal cysticercosis is caused by the taenid *T. crassiceps* and has been useful in exploring the physiological host factors associated with porcine cysticercosis, and to some degree, with human neurocysticercosis.[40] Intraperitoneal *T. crassiceps* cysticercosis of mice lends itself well to controlled and reproducible experimentation that generates numerical data of parasite loads in individual mice in a matter of weeks after infection. Its general representation of other forms of cysticercosis was later strengthened by similar results in other mouse and parasite strains, by the parasite's extensive sharing of antigens with other taennids and cestodes, and by the DNA homology between *T. crassiceps* and *T. solium*.[41] These characteristics have made murine cysticercosis a convenient instrument to test vaccine candidates and new drugs or treatments against cysticercosis.[40] Several features of natural cysticercotic disease have been found by extrapolation from experimental murine cysticercosis.[42] In *T. crassiceps* cysticercosis, females of all strains of mice studied sustain greater infection than males, though during chronic infection (more than 4 weeks) this difference disappears and male BALB/c mice show a feminization process characterized by high serum E2 (200 times the normal values) and decreased (by 90%) testosterone.[43] At the same time, the cellular immune response (Th1) is markedly diminished in both genders, while the humoral (Th2) response is

enhanced.[44] It has been reported that E2 obstructs parasite growth.[45] Gonadectomy alters this resistance pattern by increasing the intensity of infection in males and diminishing it in females,[46] although serum sex steroid levels were not detectable in these animals. However, the absence of estrogens does not prevent parasite growth in either sex, demonstrating that although E2 favors *T. crassiceps* development, it is not indispensable for rapid parasite growth.[43]

Taenia solium

Host sexual dimorphism in *T. solium* infections is much less obvious than that of experimental *T. crassiceps* cysticercosis in laboratory mice. However, women more frequently develop generalized encephalitis than do men, and when bearing subarachnoidal and ventricular vesicular parasites, women show higher inflammatory profiles in their cerebral spinal fluid than do men,[47,48] but no other immunological signs of sexual dimorphism were found.[49] It was recently reported that women harbor more single cysticercotic calcified lesions in their right cerebral hemisphere than in the left one. Presumably, lateralization of calcified cysticerci reflects the differential immunological abilities between the cerebral hemispheres.[50] Although no sexual differences in these cerebral abilities have been notified, cysticercus lateralization is not found in male neurocysticercotic patients. Sex steroids play an important role during *T. solium* infection; in particular, progesterone has been proposed to be a key immunomodulatory hormone involved in susceptibility to human taeniosis in woman and in pigs. The effect of progesterone administration upon experimental taeniosis in hamsters (*Mesocricetus auratus*) has been evaluated. It was found that progesterone-treated hamsters show a reduction in adult worm recovery by 80%, compared with both vehicle-treated and nonmanipulated infected animals. In contrast to control and vehicle groups, progesterone treatment diminished tapeworm length by 75% and increased the proliferation rate of leukocytes from spleen and mesenteric lymph nodes of infected hamsters by fivefold. IL-4, IL-6, and TNF-α expression at the duodenal mucosa promoted local exacerbation of inflammatory infiltrate, thus preventing the worm from attaching.[51] The issue of sexual dimorphism in naturally acquired porcine cysticercosis is a bit stronger than it is in human cysticercosis. Male rural pigs castrated 4 months before sacrifice show a cysticer-cosis prevalence double that of noncastrated male pigs.[52,53] Moreover, frequency of *T. solium* pig cysticercosis is increased during pregnancy, when there is a significant increase in progesterone levels.[53,54] It has also been demonstrated that castration in naturally infected male boars induces an increase in the prevalence of cysticercosis, which highlights the possible role of host androgens to restrict parasite establishment and estrogens to facilitate it.[53]

Concluding remarks

The evidence presented in this short review illustrates the role of sex steroids on immune system interactions during parasite infections, and provides clues to the many other possible mechanisms of parasite establishment, growth, and reproduction in an immunocompetent host. Furthermore, strong immunoendocrine interactions may have implications in the control of transmission and treatment of several parasitic diseases in animals, including humans. The complexity of the host–parasite relationship suggests that all physiological factors (e.g., sex, age) should be taken into consideration in the design of vaccines and new drugs. The differential response of parasites to sex steroids may also be involved in their ability to grow faster in female or male hosts.

Acknowledgments

Financial support was provided by Grant IN 214011–3 from the Programa de Apoyo a Proyectos de Investigación Científica e Innovación Tecnológica (PAPIIT), Dirección General de Asuntos del Personal Académico (DGAPA), Universidad Nacional Autónoma de México (UNAM) to J. Morales-Montor. K. Nava-Castro has a postdoctoral fellowship from DGAPA, UNAM. R. Hernández-Bello has a postdoctoral fellowship from Instituto de Ciencia y Tecnología del Distrito Federal (ICyTDF).

Conflicts of interest

The authors declare no conflicts of interest.

References

1. do Prado, J.C., Jr., A.M. Levy, M.P. Leal, *et al.* 1999. Influence of male gonadal hormones on the parasitemia and humoral response of male *Calomys callosus* infected with the Y strain of Trypanosoma cruzi. *Parasitol. Res.* **85:** 826–829.
2. Libonati, R.M., M.G. Cunha, J.M. Souza, *et al.* 2006. Estradiol, but not dehydroepiandrosterone, decreases parasitemia and increases the incidence of cerebral malaria and the mortality in plasmodium berghei ANKA-infected CBA mice. *Neuroimmunomodulation* **13:** 28–35.
3. Liesenfeld, O., T.A. Nguyen, C. Pharke & Y. Suzuki. 2001. Importance of gender and sex hormones in regulation of susceptibility of the small intestine to peroral infection with *Toxoplasma gondii* tissue cysts. *J. Parasitol.* **87:** 1491–1493.
4. Remoue, F., D. To Van, A.M. Schacht, *et al.* 2001. Gender-dependent specific immune response during chronic human *Schistosomiasis haematobia*. *Clin. Exp. Immunol.* **124:** 62–68.
5. Satoskar, A. & J. Alexander. 1995. Sex-determined susceptibility and differential IFN-gamma and TNF-alpha mRNA expression in DBA/2 mice infected with *Leishmania mexicana*. *Immunology* **84:** 1–4.
6. Vargas-Villavicencio, J.A., C. Larralde & J. Morales-Montor. 2006. Gonadectomy and progesterone treatment induce protection in murine cysticercosis. *Parasite Immunol.* **28:** 667–674.
7. Shibayama, M., Jde. Serrano-Luna, S. Rojas-Hernandez, *et al.* 2003. Interaction of secretory immunoglobulin A antibodies with Naegleria fowleri trophozoites and collagen type I. *Can. J. Microbiol.* **49:** 164–170.
8. Brodsky, F.M., L. Lem, A. Solache & E.M. Bennett. 1999. Human pathogen subversion of antigen presentation. *Immunol. Rev.* **168:** 199–215.
9. Bouman, A., M.J. Heineman & M.M. Faas. 2005. Sex hormones and the immune response in humans. *Hum. Reprod. Update* **11:** 411–423.
10. Morales-Montor, J., C. Hallal-Calleros, M.C. Romano & R.T. Damian. 2002. Inhibition of p-450 aromatase prevents feminisation and induces protection during cysticercosis. *Int. J. Parasitol.* **32:** 1379–1387.
11. Zuk, M. & K.A. McKean. 1996. Sex differences in parasite infections: patterns and processes. *Int. J. Parasitol.* **26:** 1009–1023.
12. Muñoz-Cruz, S., C. Togno-Pierce & J. Morales-Montor. 2011. Non-reproductive effects of sex steroids: their immunoregulatory role. *Curr. Top Med. Chem.* **11:** 1714–1727.
13. Bebo, B.F., Jr., A. Fyfe-Johnson, K. Adlard, *et al.* 2001. Low-dose estrogen therapy ameliorates experimental autoimmune encephalomyelitis in two different inbred mouse strains. *J. Immunol.* **166:** 2080–2089.
14. Da Silva, J.A. 1999. Sex hormones and glucocorticoids: interactions with the immune system. *Ann. N.Y. Acad. Sci.* **876:** 102–117.
15. Boulware, M.I., J.P. Weick, B.R. Becklund, *et al.* 2005. Estradiol activates group I and II metabotropic glutamate receptor signaling, leading to opposing influences on cAMP response element-binding protein. *J. Neurosci.* **25:** 5066–5078.
16. Auricchio, F., A. Migliaccio & G. Castoria. 2008. Sex-steroid hormones and EGF signalling in breast and prostate cancer cells: targeting the association of Src with steroid receptors. *Steroids* **73:** 880–884.
17. Dosiou, C., A.E. Hamilton, Y. Pang, *et al.* 2008. Expression of membrane progesterone receptors on human T lymphocytes and Jurkat cells and activation of G-proteins by progesterone. *J. Endocrinol.* **196:** 67–77.

18. Moussatche, P. & T.J. Lyons. 2012. Non-genomic proges-terone signalling and its non-canonical receptor. *Biochem. Soc. Trans.* **40:** 200–204.

19. Benten, W.P., M. Lieberherr, O. Stamm, *et al.* 1999. Testos-terone signaling through internalizable surface receptors in androgen receptor-free macrophages. *Mol. Biol. Cell* **10:** 3113–3123.

20. Gay-Andrieu, F., G.J. Cozon, J. Ferrandiz & F. Peyron. 2002. Progesterone fails to modulate *Toxoplasma gondii* replica-tion in the RAW 264.7 murine macrophage cell line. *Parasite Immunol.* **24:** 173–178.

21. Kankova, S., P. Kodym & J. Flegr. 2011. Direct evidence of Toxoplasma-induced changes in serum testosterone in mice. *Exp. Parasitol.* **128:** 181–183.

22. Pung, O.J. & M.I. Luster. 1986. *Toxoplasma gondii*: decreased resistance to infection in mice due to estrogen. *Exp. Para-sitol.* **61:** 48–56.

23. Kamis, A.B. & J.B. Ibrahim. 1989. Effects of testosterone on blood leukocytes in *Plasmodium berghei*-infected mice. *Parasitol. Res.* **75:** 611–613.

24. Rohrig, G., W.A. Maier & H.M. Seitz. 1999. Growth-stimulating influence of human chorionic gonadotropin (hCG) on *Plasmodium falciparum* in vitro. *Zentralbl. Bak-teriol.* **289:** 89–99.

25. Wunderlich, F., P. Marinovski, W.P. Benten, *et al.* 1991. Testosterone and other gonadal factor(s) restrict the effi-cacy of genes controlling resistance to *Plasmodium chabaudi* malaria. *Parasite Immunol.* **13:** 357–367.

26. Pong, C.K., A.D. Thevenon, J.A. Zhou & D.W. Taylor. 2009. Influence of human chorionic gonadotropin (hCG) on in vitro growth of *Plasmodium falciparum*. *Malar. J.* **8:** 101.

27. Klein, S.L. 2004. Hormonal and immunological mecha-nisms mediating sex differences in parasite infection. *Para-site Immunol.* **26:** 247–264.

28. Travi, B.L., Y. Osorio, P.C. Melby, *et al.* 2002. Gender is a major determinant of the clinical evolution and immune response in hamsters infected with *Leishmania spp. Infect. Immun.* **70:** 2288–2296.

29. Lezama-Davila, C.M., A.P. Isaac-Marquez, J. Barbi, *et al.* 2007. 17Beta-estradiol increases *Leishmania mexicana* killing in macrophages from DBA/2 mice by enhancing pro-duction of nitric oxide but not pro-inflammatory cytokines. *Am. J. Trop. Med. Hyg.* **76:** 1125–1127.

30. Liu, L., W.P. Benten, L. Wang, *et al.* 2005. Modulation of *Leishmania donovani* infection and cell viability by testos-terone in bone marrow-derived macrophages: signaling via surface binding sites. *Steroids* **70:** 604–614.

31. Liu, L., L. Wang, Y. Zhao, *et al.* 2006. Testosterone attenuates p38 MAPK pathway during *Leishmania donovani* infection of macrophages. *Parasitol. Res.* **99:** 189–193.

32. Pinto, A.G., L.G. Gaetano, A.M. Levy, *et al.* 2010. Experi-mental Chagas' disease in orchiectomized *Calomys callosus* infected with the CM strain of *Trypanosoma cruzi*. *Exp. Parasitol.* **124:** 147–152.

33. de Souza, E.M., M.T. Rivera, T.C. Araujo-Jorge & S.L. de Castro. 2001. Modulation induced by estradiol in the acute phase of *Trypanosoma cruzi* infection in mice. *Parasitol. Res.* **87:** 513–520.

34. dos Santos, C.D., M.P. Toldo & J.C. do Prado, Jr. 2005. *Trypanosoma cruzi*: the effects of dehydroepiandrosterone (DHEA) treatment during experimental infection. *Acta Trop.* **95:** 109–115.

35. Santos, C.D., M.P. Toldo, A.M. Levy, *et al.* 2007. Dehy-droepiandrosterone affects *Trypanosoma cruzi* tissue para-site burdens in rats. *Acta Trop.* **102:** 143–150.

36. Santos, C.D., J.C. Prado, Jr., M.P. Toldo, *et al.* 2007. *Try-panosoma cruzi*: plasma corticosterone after repetitive stress during the acute phase of infection. *Exp. Parasitol.* **117:** 405–410.

37. Santos, C.D., M.P. Toldo, F.H. Santello, *et al.* 2008. De-hydroepiandrosterone increases resistance to experimental infection by *Trypanosoma cruzi*. *Vet. Parasitol.* **153:** 238–243.

38. Kuehn, C.C., L.G. Oliveira, C.D. Santos, *et al.* 2011. Prior and concomitant dehydroepiandrosterone treatment affects immunologic response of cultured macrophages infected with *Trypanosoma cruzi* in vitro. *Vet. Parasitol.* **177:** 242–246.

39. Cervantes-Rebolledo, C., N. Moreno-Mendoza, J. Morales-Montor, *et al.* 2009. Gonadectomy inhibits development of experimental amoebic liver abscess in hamsters through downregulation of the inflammatory immune response. *Parasite Immunol.* **31:** 447–456.

40. Sciutto, E., G. Fragoso & C. Larralde. 2011. *Taenia crassiceps* as a model for *Taenia solium* and the S3Pvac vaccine. *Parasite Immunol.* **33:** 79–80.

41. Rishi, A.K. & D.P. McManus. 1988. Molecular cloning of *Taenia solium* genomic DNA and characterization of taeniid cestodes by DNA analysis. *Parasitology* **97**(Pt. 1): 161–176.

42. Morales-Montor, J., G. Escobedo, J.A. Vargas-Villavicencio & C. Larralde. 2008. The neuroimmunoendocrine network in the complex host–parasite relationship during murine cysticercosis. *Curr. Top Med. Chem.* **8:** 400–407.

43. Larralde, C., J. Morales, I. Terrazas, *et al.* 1995. Sex hor-mone changes induced by the parasite lead to feminization of the male host in murine *Taenia crassiceps* cysticercosis. *J. Steroid. Biochem. Mol. Biol.* **52:** 575–580.

44. Terrazas, L.I., R. Bojalil, T. Govezensky & C. Larralde. 1998. Shift from an early protective Th1-type immune response to a late permissive Th2-type response in murine cysticercosis (*Taenia crassiceps*). *J. Parasitol.* **84:** 74–81.

45. Terrazas, L.I., R. Bojalil, T. Govezensky & C. Larralde. 1994. A role for 17-beta-estradiol in immunoendocrine regula-tion of murine cysticercosis (*Taenia crassiceps*). *J. Parasitol.* **80:** 563–568.

46. Huerta, L., L.I. Terrazas, E. Sciutto & C. Larralde. 1992. Im-munological mediation of gonadal effects on experimental murine cysticercosis caused by *Taenia crassiceps* metaces-todes. *J. Parasitol.* **78:** 471–476.

47. Del Brutto, O.H., E. Garcia, O. Talamas & J. Sotelo. 1988. Sex-related severity of inflammation in parenchymal brain cysticercosis. *Arch. Intern. Med.* **148:** 544–546.

48. Fleury, A., A. Dessein, P.M. Preux, *et al.* 2004. Symptomatic human neurocysticercosis—age, sex and exposure factors relating with disease heterogeneity. *J. Neurol.* **251:** 830–837.

49. Chavarria, A., A. Fleury, E. Garcia, *et al.* 2005. Relationship between the clinical heterogeneity of neurocysticercosis and the immune-inflammatory profiles. *Clin. Immunol.* **116:** 271–278.

50. Meador, K.J., D.W. Loring, P.G. Ray, *et al.* 2004. Role of cerebral lateralization in control of immune processes in humans. *Ann. Neurol.* **55:** 840–844.

51. Escobedo, G., I. Camacho-Arroyo, P. Nava-Luna, *et al.* 2011. Progesterone induces mucosal immunity in a rodent model of human taeniosis by *Taenia solium. Int. J. Biol. Sci.* **7:** 1443–1456.

52. Morales, J., J.J. Martinez, J. Garcia-Castella, *et al.* 2006. *Tae-nia solium*: the complex interactions, of biological, social, geographical and commercial factors, involved in the transmission dynamics of pig cysticercosis in highly endemic areas. *Ann. Trop. Med. Parasitol.* **100:** 123–135.

53. Morales, J., T. Velasco, V. Tovar, *et al.* 2002. Castration and pregnancy of rural pigs significantly increase the prevalence of naturally acquired *Taenia solium* cysticercosis. *Vet. Parasitol.* **108:** 41–48.

54. Pena, N., J. Morales, J. Morales-Montor, *et al.* 2007. Impact of naturally acquired *Taenia solium* cysticercosis on the hormonal levels of free ranging boars. *Vet. Parasitol.* **149:** 134–137.

Ann. N.Y. Acad. Sci. ISSN 0077-8923

ANNALS OF THE NEW YORK ACADEMY OF SCIENCES
Issue: *Neuroimmunomodulation in Health and Disease*

Extrathymic CD4⁺CD8⁺ lymphocytes in Chagas disease: possible relationship with an immunoendocrine imbalance

Ana R. Pérez,[1] Alexandre Morrot,[2] Luiz R. Berbert,[3] Eugenia Terra-Granado,[3,4] and Wilson Savino[3]

[1]Institute of Immunology, Faculty of Medical Sciences, National University of Rosario, Rosario, Argentina. [2]Department of Immunology, Microbiology Institute, Federal University of Rio de Janeiro, Rio de Janeiro, Brazil. [3]Laboratory on Thymus Research, Oswaldo Cruz Institute, Oswaldo Cruz Foundation, Rio de Janeiro, Brazil. [4]Pediatric Hematology and Oncology Program, Research Center, National Cancer Institute, Rio de Janeiro, Brazil

Address for correspondence: Ana Rosa Pérez, Institute of Immunology, Faculty of Medical Sciences, National University of Rosario, Santa Fe 3100 (2000), Rosario, Argentina. perez_anarosa@yahoo.com.ar; perez.ana@conicet.gov.ar

Double-positive (DP) CD4⁺CD8⁺ T cells normally represent a thymic subpopulation that is developed in the thymus as a precursor of CD4⁺ or CD8⁺ single-positive T cells. Recent evidence has shown that DP cells with an activated phenotype can be tracked in secondary lymph organs. The detection of an activated DP population in the periphery, a population that expresses T cell receptors unselected during thymic negative selection in murine models of *Trypanosoma cruzi* infection and in humans with Chagas disease, raise new questions about the relevance of this population in the pathogenesis of this major parasitic disease and its possible link with immunoendocrine alterations.

Keywords: CD4⁺CD8⁺ double-positive CD4⁺CD8⁺ T cells; thymus; Chagas disease; *Trypanosoma cruzi* infection; cortisol; dehydroepiandrosterone

Double-positive CD4⁺CD8⁺ T cells: from physiology to pathology

The simultaneous expression of CD4 and CD8 in T cells was, until some years ago, generally considered exclusive for T cells in the thymus. However, T cells expressing both CD4 and CD8 coreceptors have been described in healthy individuals, as well as in pathological conditions such as infectious diseases, autoimmune diseases, chronic inflammatory disorders, and certain lymphoblastic diseases (Table 1).[1–3] Most double-positive (DP) CD4⁺ CD8⁺ T cells undergo differentiation to single-positive (SP) CD4⁺ or CD8⁺ mature T cells in the thymus. However, the presence of T cells coexpressing both CD4 and CD8 in the periphery raises new questions, such as, What is the origin of these lymphocytes? Do these cells exit from the thymus as immature or mature T cells? Is it possible that during the development of inflammatory responses SP T cells acquire any other cell marker, thus showing a certain degree of plasticity? Do the extrathymic DP cells have specific functions? An additional, more intriguing question is: are DP cells involved in the pathophysiology of autoimmune events occurring in some infectious conditions, particularly in the pathogenesis of Chagas disease?

Before addressing these issues, it is worthwhile to provide a general background on intrathymic DP cell differentiation and T cell exportation to the periphery of the immune system. The thymus is the primary lymphoid organ responsible for the differentiation of T cells. This process involves differential expression of CD4 or CD8 accessory molecules (among others) and rearrangements of T cell receptor (TCR) genes. The most immature thymocytes express neither the TCR complex nor the CD4 or CD8 markers, and for this reason they are called *double-negative cells*, a subset representing nearly 5% of total thymocytes. The process of maturation follows with the acquisition of CD4 and CD8 markers, generating the DP cells, which constitute

doi: 10.1111/j.1749-6632.2012.06627.x

Table 1. Diseases with presence of DP CD4$^+$CD8$^+$ T cells in periphery

Diseases with ↑% DP	Functional characteristics	Phenotype	Human or animal model	References
Infectious diseases				
Chagas disease	Citotoxic activity	Activated phenotype	Murine	27
	↑INF-γ production	↑ expression CD44 / CD69	and	
		↑ expression HLA-DR / VLA-4	human	
	↑% *T. cruzi*-specific DP cells	Activated phenotype	Human	28
	Cytotoxic activity			
	↑INF-γ production			
Malaria			Murine	32
T. evansi infection			Sheep	31
Hepatitits		↑% CD4^{+high}CD8^{+low}, CD4^{+low}CD8^{+high}, and CD4^{+high}CD8^{+high}	Human	33
		Effector/memory T cells		36
HIV	↑%HIV-specific DP cells		Human	37
	↑INF-γ production			
	↑IL-2 production			
	↑expression of cytolytic-associated lysosomal-associated membrane protein			
		↑CD8^{high}CD4^{low}	Human	38,39
SIV		↑Memory markers (CD28CD95CD45RA^{low} CD62L^{low})	Monkey	40
		↓% CCR7		
HTLV-1			Human	41
		↑ CD45RO$^+$CD18$^+$CD54$^+$	Human	42
			Human	43
HTLV-1/CMV co-infection			Human	44
HBV		↑ CD4^{low}CD8^{high}HLA-DR$^+$ ↑CD4^{high}CD8^{low}CD56$^+$CD57$^+$	Human	34
Chlamydia pneumoniae infection			Murine	45
Autoimmune diseases and other noninfectious processes				
Immunosenescence		↑% CD4^{high}CD8^{low}, CD4^{low}CD8^{high}	Human	46
Myasthenia gravis			Human	47
Sjögren's syndrome			Human	48
Multiple sclerosis			Human	49
Sclerosis	↑% DP in skin ↑production of IL-4		Human	50
Thyroiditis	DP cells in thyroid		Human	29
Rheumatoid arthritis	DP cells in synovial fluids		Human	30

Continued

Table 1. *Continued*

Diseases with ↑% DP	Functional characteristics	Phenotype	Human or animal model	References
Myelodysplastic syndromes	↓% DP		Human	51
Leukemia associated with HTLV-1 infection			Human	52
			Human	53
Breast cancer	Cytotoxic activity ↑production of IL-5 and IL-13.	Effector/memory activated CD8+	Human	54
Hodgkin lymphoma		Activated/regulatory phenotype	Human	55
		Expression of CD3+ CD5+ CD2+ CD7+ CD1a− and TDT	Human	56

75–80% of the whole thymocyte population. At this stage, TCRs are expressed on the cell surface, allowing two crucial events for thymocyte differentiation: positive and the negative selection. Differentiation of α/β TCR-expressing T cells involves an obligatory interaction with self-major histocompatibility complex molecules (MHC) in the thymus. This process, called *positive selection*, not only rescues thymocytes from programmed cell death but also induces their differentiation into mature T cells. Another critical event in thymic development is to prevent maturation of hazardous autoreactive T cells, thus eliminating T cells with self-reactive receptors (*negative selection*). Negative selection allows the establishment of self-tolerance in the T cell repertoire, promoting the apoptosis of T cells that might react against self-proteins. At this stage, immature DP thymocytes become mature SP cells, constituting nearly 20% of thymocytes. The majority of thymocytes die in the thymus, and only a small proportion leaves the organ as recent thymic emigrants (RTEs). Overall, this process is partially controlled by immunoneuroendocrine circuits.[4,5] Of note, the survival of DP cells is negatively influenced by glucocorticoids (GCs) and tumor necrosis factor-alpha (TNF-α).[6–9]

In normal conditions, it is estimated that the thymuses of adult mice export daily only 1% of total thymocytes, representing between 1 and 2 million cells.[10,11] During their journey, thymocytes interact with diverse components of the thymic microenvironment, comprising thymic epithelial cells, macrophages, dendritic cells, fibroblasts, as well as extracellular matrix (ECM) proteins, such as fibronectin or laminin. ECM-mediated interactions can influence the general process of thymocyte maturation, differentiation, and consequent T cell export to the peripheral immune system.

Are DP cells relevant in the pathophysiology of Chagas disease?

Chagas disease, a tropical neglected disease, is caused by the parasite *Trypanosoma cruzi*. Nearly 12 million people are infected in Latin America,[12] and it has spread to nonendemic zones, including the United States, Europe, Asia, and Oceania, and represents a new world health problem, considering that Chagas disease can also be spread by congenital transmission, blood transfusion, or organ transplantation.[13] The clinical manifestations of disease can be highly heterogeneous. After the first contact with the parasite, infected individuals develop an acute phase of disease, which can be asymptomatic or present fever, lymphoadeno, and/or splenomegaly. Following the disappearance of the parasite in the blood, infected individuals can remain asymptomatic for the rest of their lives, a period known as the "indeterminate form of the chronic phase of disease." Between 15 and 30 years after the initial infection, nearly 30% of infected individuals develop chronic chagasic myocarditis, a hallmark of disease. The causes and mechanisms associated with the development and the establishment of different clinical manifestations of Chagas disease seem complex and remain to be precisely defined. Yet, autoimmune reactions, as well as the consequences of parasite persistence, have already been largely studied.[14,15] It is possible that the indeterminate form of chronic Chagas disease occurs when immunological response of the host against the parasite is more efficient, as during the acute phase, whereas the symptomatic forms seen in the chronic phase seem to occur in patients with hyperergic or inefficient immune responses.[16]

SP T cells play a key role in both the protection against the parasite and in immunopathology.[17–19] For this reason, historically, the study of Chagas disease has mainly focused on the effector immune response directed at the parasite, as well as autoimmune reactions observed against the heart. Nevertheless, the putative involvement of the thymus-derived T lymphocytes in the immunopathology of Chagas disease, as well as the impact of the host's response in the thymus, have not been explored. In this respect, we recently found that RTEs in acutely infected chagasic animals are essentially SP cells, and that their relative and absolute numbers increase progressively with the infection, suggesting an abnormal release of T cells.

During *T. cruzi* infection, the thymus is severely affected by a dysregulated circuit of proinflammatory cytokines and hypothalamus–pituitary–adrenal (HPA)-related hormones.[20,21] Thymus atrophy is commonly observed during murine models of *T. cruzi* acute infection and persists, although to a lower extent, during the chronic phase. Thymic atrophy is essentially secondary to massive depletion of the DP T cell population, at least partially caused by an immunoendocrine imbalance, with enhanced levels of GCs as a result of an exaggerated increase in the levels of inflammatory cytokines.[21,22] In addition to the apoptosis seen in the DP subset, other changes might account for the loss of these cells, such as an increase in their export from the thymus, a decrease in their proliferation rate, and/or a diminution in their numbers secondary to a low recruitment of bone marrow–derived precursors.[23] Among the causes that could influence an increase in the DP cell exportation is the abnormally high intrathymic expression of ECM ligands and receptors.[24]

The first evidence of an aberrant thymic release of DP lymphocytes to the periphery after murine acute *T. cruzi* infection was the observation that DP cells progressively accumulated in peripheral lymphoid organs.[24] Actually, they can also be detected in circulation and in low numbers within the heart, a major target organ in Chagas disease (Fig. 1A). The DP cells seem to be thymus-dependent because their presence in the periphery is largely reduced when the infection is carried out in thymectomized mice. Moreover, studies performed in BALB/c mice showed that some of these extrathymic DP cells carry prohibited V_β segments of the TCR,[24,25] leading to the hypothesis that they have escaped from negative selection, and thus have the potential to be autoreactive.

As seen in Figure 1B, an abnormal increase in DP cell export results in a progressive augmentation in their relative and absolute numbers in both lymph nodes and the spleen of infected animals. In these studies, thymocytes were previously intrathymically labeled with fluorescein isothiocyanate (FITC), and 16 hours later RTEs (FITC+ cells) were tracked in peripheral lymphoid organs.[23] The augmented presence of DP cells in peripheral organs (including the heart) might represent an accelerated recruitment of T cells from thymus as a compensatory mechanism to overcome the anergy/immunosupression described during the acute phase of *T. cruzi* infection.[26]

More recently, we determined that such abnormal extrathymic DP cells bear an activated phenotype, with upregulated expression of the activation markers CD44 and CD69 at levels comparable with activated/peripheral CD4+ or CD8+ SP T cells. Interestingly, CD62L (L-selectin) is expressed normally in RTEs and mature SP T cells; but during acute infection, CD62L expression in DP cells within the thymus is comparable to expression of CD62L on SP cells in the periphery.[27] In fact, upregulation of CD62L, which directs lymphocyte homing to lymph nodes, may favor the exit of DP cells from the thymus of *T. cruzi*–infected individuals.

In addition to the activated status of T cells, peripheral DP cells upregulate TCR expression levels (Fig. 1C). Surface TCR expression level greatly influences T cell antigen sensitivity. This level is rapidly downregulated when T cells are stimulated with strong TCR agonists, as engaged TCRs are internalized and degraded. The increased expression of TCR on peripheral DP cells may help promote sustained antigenic signaling in the

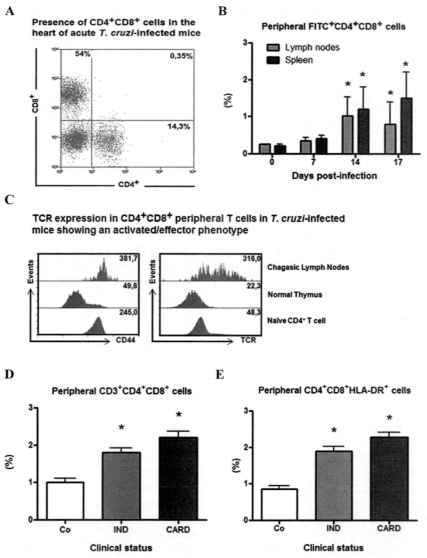

Figure 1. Intrathymic and extrathymic CD4⁺CD8⁺ T cells during experimental and human *T. cruzi* infection. (A) Representative dot plot showing the presence of CD4⁺CD8⁺ cells in the heart during acute *T. cruzi* infection. CD4⁺CD8⁺ cells were not detectable in control mice. (B) The proportions of immature thymus-derived CD4⁺CD8⁺ lymphocytes progressively increase with the course of infection in both lymph nodes and spleen in *T. cruzi*–infected mice. Control and *T. cruzi* acutely infected mice received intrathymic injection of FITC, and 16 hours later were analyzed by flow cytometry to detect CD4⁺CD8⁺FITC⁺ as recent thymic emigrant cells in peripheral lymphoid organs. (C) DP cells show an activated/effector phenotype in the periphery of *T. cruzi*–infected mice with high levels of TCR expression. Lymphocytes were isolated from the thymus, subcutaneous lymph nodes, and spleen, during acute phase. Representative histograms of CD44 (left panel); TCR expression levels in CD4⁺CD8⁺ T cells from chagasic lymph nodes and normal thymus, and naive CD4⁺ T cells from noninfected mice as a control (right panel). The values in the upper-right corner indicate the mean fluorescence intensity from the expression of the markers in each histogram. Differences between chagasic DP from peripheral lymph nodes *versus* normal thymic DP cells and naive T cells from the spleen are significant ($P < 0.05$). (D) Proportion of peripheral blood CD4⁺CD8⁺ cell subset within CD3⁺ T lymphocytes from healthy human individuals (Co), asymptomatic/indeterminate (IND), or with myocardiopathy (CARD) chronic chagasic patients. Cells were analyzed by flow cytometry ($n = 13$–15 individuals per group). *$P < 0.05$ versus Co individuals. (E) Proportion of peripheral blood CD4⁺CD8⁺ HLA-DR⁺ cell subset on CD3⁺ T lymphocytes from healthy individuals (Co), asymptomatic/indeterminate (IND), or with myocardiopathy (CARD) chronic chagasic patients. Cells were analyzed by flow cytometry ($n = 13$–15 individuals per group). *$P < 0.05$ versus Co individuals.

activation pathway of these cells during *T. cruzi* infection. With an extended period of antigenic stimulation, peripheral DP lymphocytes could promptly reach the threshold of activation required to gain effector/memory functions or, alternatively, to further differentiate into antigen-specific SP cells, with a possible role in cell-mediated immunoprotection.[3] In fact, peripheral DP cells from infected mice have been shown to produce high levels of IFN-γ mRNA.[27] Furthermore, our results indicate that DP cells purified from peripheral lymphoid tissues of infected animals show a cytotoxic capacity comparable to that of naive SP T cells.[27]

As the experimental models indicate a premature release of immature DP thymocytes in both acute and chronic experimental Chagas disease,[25] it was plausible to address whether this phenomenon also occurs in chagasic patients. To evaluate this, we examined the frequency of peripheral blood DP cells in both chronic chagasic patients at the indeterminate phase of disease and in individuals with chronic myocarditis. The results showed a higher percentage of DP cells in cardiac chagasic patients compared with healthy individuals (Fig. 1D). Of most relevance, we found that patients with the cardiac form of Chagas disease presented with higher percentages of peripheral blood HLA-DR⁺ (MHC class II) DP cells, compared with noninfected individuals (Fig. 1E). More recently, similar results were reported by an independent research group.[28] Of note, Giraldo *et al.* showed that human DP T cells can also recognize a parasite-derived MHC class I epitope during chronic infection, probably contributing to *T. cruzi*–induced cardiopathy.[28]

In addition, we phenotyped extrathymic DP T cells for the expression of VLA-4, an integrin-type receptor of fibronectin that also can be seen as a T cell activation marker.[27] Our findings indicate that the increased percentages of circulating DP cells exhibit a fully activated HLA-DR^high/VLA-4^high pattern, which is enhanced in patients with the severe cardiac form of chronic Chagas disease.[27] Blood DP cells obtained from infected hosts showed marked functional plasticity, exerting both cytolytic activity and/or Th1 helper activity.[27]

As mentioned previously, *T. cruzi* acute infection results in severe thymic atrophy and an early release of DP T cells with autoreactive TCRs into the periphery.[24] We recently showed that despite the thymic atrophy, the machinery necessary to achieve negative selection remains functional during the acute phase of infection,[27] suggesting that DP cells are shunted, escaping the checkpoints required for maturation of T cells and that normally eliminate T cells that express "forbidden" TCRs. Regardless of the specificity of DP T cells, the expression of both CD4 and CD8 in extrathymic DP cells may diminish the functional threshold for antigen-MHC–specific T cell recognition, and/or decrease the requirements of costimulatory signals provided by antigen-presenting cells (APCs) to generate T cell activation. This activation could induce cytotoxic activity, not only against parasite antigens but also toward cells bearing self-antigens, leading to the formation of cross-reactive or neoepitopes, and thus favoring autoimmune reactions in the target tissues.

Further studies are required to elucidate the mechanisms that allow the emergence of potentially autoreactive DP cells in Chagas disease. Nevertheless, we cannot rule out that SP cells in the periphery can become DP cells after a prolonged antigenic stimulation, which could occur as a result of the persistence of the parasite. We might speculate that SP cells become self-reactive after expression of the other coreceptor, thereby inducing an increase in the affinity of the TCR to self-antigens. However, whatever their origin, the data indicate that activated extrathymic DP T cells might be associated with the development of the cardiac clinical form of the disease.

Of note, several other infectious and autoimmune diseases are characterized by the presence of circulating DP T cells, as summarized in Table 1. Interestingly, the detection of DP T cells in the target organs of some autoimmune diseases, such as thyroiditis or rheumatoid arthritis, suggests a connection with the pathophysiology of the corresponding disease.[29,30] Moreover, extrathymic DP T lymphocytes have been detected in other parasitic diseases, including infections by *Trypanosoma evansi* or *Plasmodium berghei*,[31,32] suggesting that these atypical T cells may represent a compensatory response of the immune system to cope with the pathogens. The fact that extrathymic DP cells during *T. cruzi* or viral infections present with an activated phenotype, and thus might exert cytotoxic activity, supports this notion.[27,28,33,34] Alternatively, it is conceivable that activated DP cells may play a

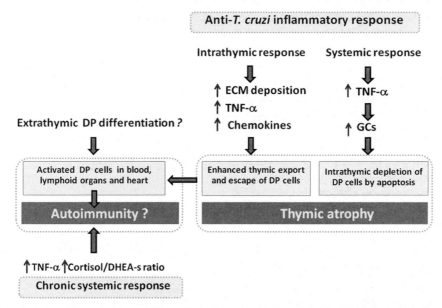

Figure 2. Hypothetical model for the impact of *T. cruzi* infection on DP T cells and their possible role in the immunopathology of Chagas disease. DP CD4+CD8+ thymocyte subset is severely affected by a dysregulated immunoendocrine circuit of proinflammatory cytokines (mainly TNF-α) and glucocorticoid hormones (GCs). Thymic atrophy is caused by apoptosis of DP thymocytes induced by GCs, although TNF-α together with extracellular matrix (ECM), such as fibronectin and laminin, combined with chemokines, may contribute to enhanced exit of cells from the organ. The abnormal thymic release of DP lymphocytes, in addition to a possible DP differentiation *de novo* in the periphery, may be related with the autoimmune component of Chagas disease. In a chronic scenario, enhanced and sustained levels of TNF-α and enhanced cortisol/DHEA-s ratio may favor the activated status of DP cells and the inflammatory events related with the autoimmune component of disease. The scheme shown in the figure was based on results obtained from studies carried out in C57BL/6 mice infected with the Tulahuen strain and also from human studies.

relevant role in modulating the skewing of adaptive immune responses via cytokine secretion. Accordingly, the cytokines secreted by these cells may address the function of APCs during the early adaptive immune responses, thus providing a link between innate and adaptive immunity. Accordingly, IFN-γ production by DP cells may be involved in the induction of protective/autoreactive Th1 cells.

Table 2. Spearman correlation analysis between percentage of double-positive (DP) CD4+CD8+ T cells and systemic cytokine and glucocorticoid hormone levels in healthy controls and chagasic patients classified as indeterminate (IND) and with cardiac involvement (CARD)

Pair correlation	Overall ($n = 39$)		Controls ($n = 10$)		IND patients ($n = 10$)		CARD patients ($n = 19$)	
	P value	r_s	*P* value	r_s	*P* value	r_s	*P* value	r_s
DPs (%) vs. TNF-α (pg/mL)	**0.0159***	0.40	0.1328	−0.54	0.8123	−0.09	**0.0374***	0.48
DPs (%) vs. IFN-γ (pg/mL)	0.1273	0.25	0.4630	−0.27	0.6821	0.14	0.5835	0.13
DPs (%) vs. Cortisol	0.6012	0.08	0.6436	−0.17	0.8916	0.05	0.8576	−0.04
DPs (%) vs. DHEA-s	**0.0408***	−0.34	**0.0083***	0.82	0.6073	−0.18	**0.0282***	−0.51
DPs (%) vs. Cortisol/DHEA-s	**0.0196***	0.38	**0.0045***	−0.85	0.4483	0.26	**0.0286***	0.51

*P values with statistical significance. Statistical significance was set up at $P < 0.05$.

NOTE: The nonparametric Spearman correlation test was performed using GraphPad Prism version 5.00 for Windows, GraphPad Software, San Diego, CA www.graphpad.com.

As activated DP cells may hypothetically recognize both class I and class II MHC, they could eliminate activated APCs, favoring parasite persistence during infection.[27]

As discussed previously, although studies in experimental models revealed that during acute infection by *T. cruzi* DP cells come from the thymus, one question that remains elusive is whether DP cells in chronic chagasic patients are thymus-derived or peripherally differentiated, or both. Future studies analyzing T cell receptor exclusion circles in circulating T cells, together with CD45RA/CD45RO markers, should help in further understanding this issue.

DP cells and the immunoendocrine imbalance in human Chagas disease

We have previously shown that a systemic inflammatory milieu, with enhanced levels of TNF-α and IFN-γ, is evident in chagasic patients with severe myocarditis, compared with to healthy subjects.[35] This scenario was paralleled by a disrupted activation of the HPA axis, characterized by decreased concentrations of dehydroepiandrosterone-sulphate (DHEAS) and an unbalanced cortisol/DHEAS ratio, reinforcing the view that severe human Chagas disease is devoid of an adequate anti-inflammatory environment, thus favoring pathology.[35] In keeping with this notion, we observed that extrathymic DP T cells positively correlated with circulating levels of TNF-α and with the cortisol/DHEAS ratio in an overall study population, and within those chagasic patients having cardiopathy (Table 2). By contrast, a negative correlation between extrathymic DP T cells and DHEAS was found in the overall population and also in cardiac patients (Table 2). This raises the question of whether there is a cause/effect relationship between immunoendocrine abnormalities and the levels of circulating extrathymic DP T cells linked to clinical progression in humans. Chronic stimulation of proinflammatory cytokines may influence the activate state of DP T cells, and at the same time, the cortisol/DHEAS ratio imbalance acts as a permissive scenario to myocarditis development. Further investigation of the relationship among TNF-α, DHEAS, and the cortisol/DHEAS ratio with the phenotype and effector functions of extrathymic DP cells, may help to elucidate the contribution of these cells to the etiology of chagasic carditis.

Conclusions

Overall, the findings support the notion that thymic alterations and the resulting accumulation of extrathymic DP cells during chagasic infection should not be simplistically viewed as a bystander phenomenon but as a relevant pathophysiological component in the course of the disease. Moreover, preserved thymus homeostasis during infections may be relevant for the development of an effective immune response. In this regard, the loss of DP thymocytes during *T. cruzi* infection, as a consequence of immunoendocrine imbalance causing cell death and abnormal exit and peripheral increase of activated DP cells, may also have an impact on tissue damage, thus contributing to the immunopathological events seen during chronic infection. Indeed, an altered cortisol/DHEAS ratio may act as a permissive hormonal environment for the maintenance of the activated state observed in DP cells from chronic chagasic patients. A hypothetical model for the possible role of DP T cells in the immunopathology of Chagas disease is shown in Figure 2. These issues represent an open field for further investigations of Chagas disease.

Acknowledgments

This work was partially funded with grants from CONICET (PIP CONICET 0789) and SeCyT-UNR (MED 244 and MED 245) from Argentina, as well as from CNPq, Faperj, and Oswaldo Cruz Foundation from Brazil, and Amsud/Pasteur network.

Conflicts of interest

The authors declare no conflicts of interest.

References

1. Parel, Y. & C. Chizzolini. 2004. CD4+CD8+ double positive (DP) T cells in health and disease. *Autoimmun. Rev.* **3:** 215–220.
2. Antica, M. & R. Scollay. 1999. Development of T lymphocytes at extrathymic sites. *J. Immunol.* **163:** 206–211.
3. Zuckermann, F.A. 1999. Extrathymic CD4/CD8 double positive T cells. *Vet. Immunol. Immunopathol.* **72:** 55–66.
4. Savino, W. & M. Dardenne. 2000. Neuroendocrine control of thymus physiology. *Endocr. Rev.* **21:** 412–443.
5. Savino, W., E. Arzt & M. Dardenne. 1999. Immunoneuroendocrine connectivity: the paradigm of the thymus-hypothalamus/pituitary axis. *Neuroimmunomodulation* **6:** 126–136.
6. Jondal, M., A. Pazirandeh & S. Okret. 2004. Different roles for glucocorticoids in thymocyte homeostasis? *Trends Immunol.* **25:** 595–600.

7. Ashwell, J.D., F.W. Lu & M.S. Vacchio. 2000. Glucocorticoids in T cell development and function. *Annu. Rev. Immunol.* **18:** 309–345.

8. Baseta, J.G. & O. Stutman. 2000. TNF regulates thymocyte production by apoptosis and proliferation of the triple negative (CD3⁻CD4⁻CD8⁻) subset. *J. Immunol.* **165:** 5621–5630.

9. Guevara Patiño, J.A., M.W. Marino, V.N. Ivanov & J. Nikolich-Zugich. 2000. Sex steroids induce apoptosis of CD8⁺CD4⁺ double-positive thymocytes via TNF-alpha. *Eur. J. Immunol.* **30:** 2586–2592.

10. Scollay, R. & D.I. Godfrey. 1995. Thymic emigration: conveyor belts or lucky dips? *Immunol. Today.* **16:** 268–273.

11. Berzins, S.P., A.P. Uldrich, J.S. Sutherland, *et al.* 2002. Thymic regeneration: teaching old immune system new tricks. *Trends Mol. Med.* **8:** 469–476.

12. World Health Organization (WHO), Pan American Health Organization (PAHO) Regional Office. 2007. The burden of neglected diseases in Latin America and the Caribbean compared with some other communicable diseases. URL http://www.paho.org/English/AD/DPC/CD/parasit.htm.

13. Coura, J.R. & P.A. Viñas. 2010. Chagas disease: a new worldwide challenge. *Nature* **465:** S6–S7.

14. Cunha-Neto, E., A.M. Bilate, K.V. Hyland, *et al.* 2006. Induction of cardiac autoimmunity in Chagas heart disease: a case for molecular mimicry. *Autoimmunity* **459:** 41–54.

15. Tarleton, R.L. 2001. Parasite persistence in the aetiology of Chagas disease. *Int. J. Parasitol.* **31:** 550–554.

16. Higuchi, M.L. 1997. Chronic chagasic cardiopathy: the product of a turbulent host-parasite relationship. *Rev. Inst. Med. Trop.* **39:** 53–60.

17. Zhang, L. & R.L. Tarleton. 1996. Characterization of cytokine production in murine *Trypanosoma cruzi* infection by in situ immunocytochemistry: lack of association between susceptibility and type 2 cytokine production. *Eur. J. Immunol.* **26:** 102–109.

18. Tarleton, R.L., J. Sun, L. Zhang & M. Postan. 1994. Depletion of T-cell subpopulations results in exacerbation of myocarditis and parasitism in experimental Chagas' disease. *Infect. Immun.* **62:** 1820–1829.

19. Marin-Neto, J.A., E. Cunha-Neto, B.C. Maciel & M.V. Simões. 2007. Pathogenesis of chronic Chagas heart disease. *Circulation* **115:** 1109–1123.

20. Corrêa-de-Santana, E., M. Paez-Pereda, M. Theodoropoulou, *et al.* 2006. Hypothalamus-pituitary-adrenal axis during *Trypanosoma cruzi* acute infection in mice. *J. Neuroimmunol.* **173:** 12–22.

21. Pérez, A.R., E. Roggero, A. Nicora, *et al.* 2007. Thymus atrophy during *Trypanosoma cruzi* infection is caused by an immuno-endocrine imbalance. *Brain Behav. Immun.* **21:** 890–900.

22. Roggero, E., A.R. Pérez, M. Tamae-Kakazu, *et al.* 2006. Endogenous glucocorticoids cause thymus atrophy but are protective during acute *Trypanosoma cruzi* infection. *J. Endocrinol.* **190:** 495–503.

23. Savino, W., D.M. Villa-Verde, D.A. Mendes-da-Cruz, *et al.* 2007. Cytokines and cell adhesion receptors in the regulation of immunity to *Trypanosoma cruzi*. *Cytokine Growth Factor Rev.* **18:** 107–124.

24. Cotta-de-Almeida, V., A. Bonomo, D.A. Mendes-da-Cruz, *et al.* 2003. *Trypanosoma cruzi* infection modulates intrathymic contents of extracellular matrix ligands and receptors and alters thymocyte migration. *Eur. J. Immunol.* **33:** 2439–2448.

25. Mendes-da-Cruz, D.A., J. de Meis, V. Cotta-de-Almeida & W. Savino. 2003. Experimental *Trypanosoma cruzi* infection alters the shaping of the central and peripheral T-cell repertoire. *Microbes Infect.* **5:** 825–832.

26. Argibay, P.F., J.M. Di Noia, A. Hidalgo, *et al.* 2002. *Trypanosoma cruzi* surface mucin TcMuc-e2 expressed on higher eukaryotic cells induces human T cell anergy, which is reversible. *Glycobiology* **12:** 25–32.

27. Morrot, A., E. Terra-Granado, A.R. Pérez, *et al.* 2011. Chagasic thymic atrophy does not affect negative selection but results in the export of activated CD4⁺CD8⁺ T cells in severe forms of human disease. *PLoS Negl. Trop. Dis.* **5:** e1268.

28. Giraldo, N.A., N.I. Bolaños, A. Cuella, *et al.* 2011. Increased CD4⁺/CD8⁺ double-positive T cells in chronic Chagasic patients. *PLoS Negl. Trop. Dis.* **5:** e1294.

29. Iwatani, Y., Y. Hidaka, F. Matsuzuka, *et al.* 1993. Intrathyroidal lymphocyte subsets, including unusual CD4⁺CD8⁺ cells and CD3ˡᵒTCR alpha-betaˡᵒ⁻CD4–CD8- cells, in autoimmune thyroid disease. *Clin. Exp. Immunol.* **93:** 430–436.

30. De Maria, A., M. Malnati, A. Moretta, *et al.* 1987. CD3⁺4⁻8⁻ WT31⁻ (T cell receptor gamma⁺) cells and other unusual phenotypes are frequently detected among spontaneously interleukin 2-responsive T lymphocytes present in the joint fluid in juvenile rheumatoid arthritis. A clonal analysis. *Eur. J. Immunol.* **17:** 1815–1819.

31. Onah, D.N., J. Hopkin & A.G. Luckins. 1998. Induction of CD4⁺CD8⁺ double positive T cells and increase in CD5⁺ B cells in efferent lymph in sheep infected with *Trypanosoma evansi*. *Parasite Immunol.* **20:** 121–134.

32. Francelin, C., L.C. Paulino, J. Gameiro & L. Verinaud. 2011. Effects of *Plasmodium berghei* on thymus: high levels of apoptosis and premature egress of CD4⁽⁺⁾CD8⁽⁺⁾ thymocytes in experimentally infected mice. *Immunobiol.* **216:** 1148–1154.

33. Nascimbeni, M., S. Pol & B. Saunier. 2011. Distinct CD4⁺CD8⁺ double-positive T cells in the blood and liver of patients during chronic hepatitis B and C. *PLoS ONE.* **6:** e20145.

34. Ortolani, C., E. Forti, E. Radin, *et al.* 1993. Cytofluorimetric identification of two populations of double positive (CD4⁺CD8⁺) T lymphocytes in human peripheral blood. *Biochem. Biophys. Res. Commun.* **191:** 601–609.

35. Pérez, A.R., S.D. Silva-Barbosa, L.R. Berbert, *et al.* 2011. Immunoneuroendocrine alterations in patients with progressive forms of chronic Chagas disease. *J. Neuroimmunol.* **235:** 84–90.

36. Nascimbeni, M., E.C. Shin, L. Chiriboga, *et al.* 2004. Peripheral CD4⁽⁺⁾CD8⁽⁺⁾ T cells are differentiated effector memory cells with antiviral functions. *Blood* **104:** 478–486.

37. Howe, R., S. Dillon, L. Rogers, *et al.* 2009. Phenotypic and functional characterization of HIV-1-specific CD4⁺CD8⁺ double-positive T cells in early and chronic HIV-1 infection. *J. Acquir. Immune Defic. Syndr.* **50:** 444–456.

38. Hughes, G.J., A. Cochrane, C. Leen, *et al.* 2008. HIV-1-infected CD8⁺CD4⁺ T cells decay in vivo at a similar rate to infected CD4 T cells during HAART. *AIDS* **22:** 57–65.

39. Ribrag, V., D. Salmon, F. Picard, *et al.* 1993. Increase in double-positive CD4⁺CD8⁺ peripheral T-cell subsets in an HIV-infected patient. *AIDS* **7:** 1530.

40. Wang, X., A. Das, A.A. Lackner, *et al.* 2008. Intestinal double-positive CD4⁺CD8⁺ T cells of neonatal rhesus macaques are proliferating, activated memory cells and primary targets for SIVMAC251 infection. *Blood* **112:** 4981–4990.

41. Furukawa, S., K. Sasai, J. Matsubara, *et al.* 1992. Increase in T cells expressing the gamma/delta receptor and CD4⁺CD8⁺ double-positive T cells in primary immunodeficiency complicated by human T-cell lymphotropic virus type I infection. *Blood* **80:** 3253–3255.

42. Macchi, B., G. Graziani, J. Zhang & A. Mastino. 1993. Emergence of double-positive CD4/CD8 cells from adult peripheral blood mononuclear cells infected with human T cell leukemia virus type I (HTLV-I). *Cell Immunol.* **149:** 376–389.

43. Ciminale, V., M. Hatziyanni, B.K. Felber, *et al.* 2000. Unusual CD4⁺CD8⁺ phenotype in a greek patient diagnosed with adult T-cell leukemia positive for human T-cell leukemia virus type I (HTLV-I). *Leuk. Res.* **24:** 353–358.

44. Ohata, J., M. Matsuoka, T. Yamashita, *et al.* 1999. CD4/CD8 double-positive adult T cell leukemia with preceding cytomegaloviral gastroenterocolitis. *Int. J. Hematol.* **69:** 92–95.

45. Penttilä, J.M., R. Pyhälä, M. Sarvas & N. Rautonen. 1998. Expansion of a novel pulmonary CD3⁽⁻⁾CD4⁽⁺⁾CD8⁽⁺⁾ cell population in mice during *Chlamydia pneumoniae* infection. *Infect. Immun.* **66:** 3290–3294.

46. Ghia, P., G. Prato, S. Stella, *et al.* 2007. Age-dependent accumulation of monoclonal CD4⁺CD8⁺ double positive T lymphocytes in the peripheral blood of the elderly. *Br. J. Haematol.* **139:** 780–790.

47. Schlesinger, I., R. Rabinowitz, T. Brenne, *et al.* 1992. Changes in lymphocyte subsets in myasthenia gravis: correlation with level of antibodies to acetylcholine receptor and age of patient. *Neurol.* **42:** 2153–2157.

48. Ferraccioli, G.F., E. Tonutti, L. Casatta, *et al.* 1996. CD4 cytopenia and occasional expansion of CD4⁺CD8⁺ lymphocytes in Sjögren's syndrome. *Clin. Exp. Rheumatol.* **14:** 125–130.

49. Ziaber, J., D. Stopczy, H. Tchórzewski, *et al.* 2000. Increased percentage of "double phenotype" form T lymphocytes in blood of patients with multiple sclerosis. *Neurol. Neurochir. Pol.* **34:** 1137–1143.

50. Parel, Y., M. Aurrand-Lions, A. Scheja, *et al.* 2007. Presence of CD4⁺CD8⁺ double-positive T cells with very high interleukin-4 production potential in lesional skin of patients with systemic sclerosis. *Arthritis Rheum.* **56:** 3459–3467.

51. Fozza, C., S. Contini, P. Virdis, *et al.* 2012. Patients with myelodysplastic syndromes show reduced frequencies of CD4⁽⁺⁾CD8⁽⁺⁾ double-positive T cells. *Eur. J. Haematol.* **88:** 89–90.

52. Raza, S., S. Naik, V.P. Kancharla, *et al.* 2010. Dual-positive (CD4⁺/CD8⁺) acute adult T-cell leukemia/lymphoma associated with complex karyotype and refractory hypercalcemia: case report and literature. *Case Rep. Oncol.* **3:** 489–494.

53. Kamihira, S., H. Sohda, S. Atogami, *et al.* 1993. Unusual morphological features of adult T-cell leukemia cells with aberrant immunophenotype. *Leuk. Lymphoma* **12:** 123–130.

54. Desfrançois, J., L. Derré, M. Corvaisier, *et al.* 2009. Increased frequency of nonconventional double positive CD4CD8 alpha-beta T cells in human breast pleural effusions. *Int. J. Cancer* **125:** 374–380.

55. Rahemtullah, A., N.L. Harris, M.E. Dorn, *et al.* 2008. Beyond the lymphocyte predominant cell: CD4⁺CD8⁺ T-cells in nodular lymphocyte predominant Hodgkin lymphoma. *Leuk. Lymphoma* **49:** 1870–1878.

56. Rahemtullah, A., K.K. Reichard, F.I. Preffer, *et al.* 2006. A double-positive CD4⁺CD8⁺ T-cell population is commonly found in nodular lymphocyte predominant Hodgkin lymphoma. *Am. J. Clin. Pathol.* **126:** 805–814.

Ann. N.Y. Acad. Sci. ISSN 0077-8923

ANNALS OF THE NEW YORK ACADEMY OF SCIENCES
Issue: *Neuroimmunomodulation in Health and Disease*

Different peripheral neuroendocrine responses to *Trypanosoma cruzi* infection in mice lacking adaptive immunity

Eduardo Roggero,[1] Johannes Wildmann,[2] Marcelo O. Passerini,[1] Adriana del Rey,[2] and Hugo O. Besedovsky[2]

[1]CAECHIS, Universidad Abierta Interamericana, Rosario, Argentina. [2]Department of Immunophysiology, Institute of Physiology and Pathophysiology, Medical Faculty, Marburg, Germany

Address for correspondence: Hugo O. Besedovsky, Department of Immunophysiology, Institute of Physiology and Pathophysiology, Deutschhausstrasse 2, 35037 Marburg, Germany. besedovs@mailer.uni-marburg.de

Trypanosoma cruzi infection in mice triggers neuroendocrine responses that affect the course of the disease. To analyze the contribution of adaptive immunity to these responses, comparative studies between normal C57Bl/6J and recombinase activator gene 1 (RAG-1)–deficient mice, which lack mature B and T lymphocytes, were performed. There was no difference between both types of mice in basal body weight. Following infection, higher parasitemia, increased IL-1β and IL-6 blood levels, less marked changes in lymphoid organs weight, no cardiomegaly, and earlier mortality were observed in RAG-1–deficient, compared with normal mice. The response of the hypothalamus–pituitary–adrenal axis after infection occurred earlier and was more intense in RAG-1–deficient mice than in normal mice. Noradrenaline concentration and serotonergic metabolism in the spleen, lymph nodes, and heart differed between RAG-1–deficient and normal mice. Our studies indicate that the absence of adaptive immunity to *T. cruzi* influences the neuroendocrine response to the infection with this parasite.

Keywords: recombinase activator gene 1–deficient mice; *Trypanosoma cruzi*; infection; corticosterone; noradrenaline; serotonin; sympathetic nervous system

Introduction

Natural and adaptive immune responses can both trigger cytokine-mediated neuroendocrine responses capable of affecting the course of infectious diseases. For example, intense stimulation of the hypothalamus–pituitary–adrenal (HPA) axis is detected in mice during the course of acute infection with *Trypanosoma cruzi*.[1] The increase in glucocorticoid levels is responsible for the thymus atrophy observed during infection with this parasite, but is protective for the host. Indeed, there is an uncontrolled release of proinflammatory cytokines during the course of the infection and early mortality when the effect of the increased levels of corticosterone is abrogated by adrenalectomy or by administration of the glucocorticoid receptor blocker RU486.[1] This evidence indicates that coupled immune and neuroendocrine responses are triggered during the infection with *T. cruzi*, the parasite that causes Chagas disease. This disease, also called American trypanosomiasis, is the fourth leading cause of death in Latin America. Although once confined to South and Central America, where it is endemic, it has now spread to other continents, particularly Europe. Indeed, over 120 million people are at risk of infection in 21 countries. In humans, the disease occurs in two stages: an acute phase, manifested shortly after infection, and a chronic state that may develop over 10 years. Chronic infection results in various neurological disorders, damage to the heart muscle (cardiomyopathy, the most serious manifestation), and sometimes dilation of the digestive tract (megacolon and megaesophagus). Left untreated, Chagas disease can be fatal, in most cases due to the cardiac sequelae.

doi: 10.1111/j.1749-6632.2012.06645.x

The innate immune system has the ability to sense pathogens via germ line–encoded pattern recognition receptors. These receptors recognize pathogen associated molecular patterns (PAMPs), conserved molecules shared by several microorganisms, including *T. cruzi*, and trigger the activation of host innate responses.[2–4] Adaptive immunity to *T. cruzi* has been well characterized, with critical involvement of CD4[+] Th1 and CD8[+] T cells that recognize *T. cruzi*–specific antigens.[2] However, it is still unknown to what extent the absence of adaptive immunity influences neuroendorine responses triggered by *T. cruzi* inoculation. To analyze whether combined T and B cell deficiency influences the response of the HPA axis and the sympathetic nervous system (SNS) and affects peripheral serotonergic mechanisms during *T. cruzi* infection, we performed comparative studies between normal C57Bl/6J (C57) mice and the recombinase activator gene 1 (RAG-1)–deficient mutant mice in the same genetic background. RAG-1–deficient mice lack mature B and T lymphocytes due to their incapacity to rearrange and recombine the immunoglobulin and T cell receptor genes.

Material and methods

Mice and infection

C57Bl/6J (C57) and RAG-1–deficient mice (original breeding pairs were obtained from Jackson Laboratory, Bar Harbor, ME) were bred in the animal facilities of the Universidad Abierta Interamericana, Rosario, Argentina. Mice were housed individually for 1 week before experiments were started and kept single-caged throughout the experiments in temperature-, humidity- and light (12-h cycles)-controlled rooms. One hundred trypomastigotes suspended in 100 µL physiological saline were injected s.c. (50 µL in each flank) when mice were 8–10 weeks old. The Tulahuén strain of *T. cruzi* used in this study was maintained by serial passages in C57BL/6 suckling mice. Eighteen days postinfection (p.i.), groups of C57 and RAG-1–deficient mice were sacrificed, and blood obtained. The spleen, the inguinal lymph nodes, and the heart were collected and weighed. Results are expressed as relative to the body weight (organ weight/body weight × 100).

Survival

A separate group of C57 and RAG-1–deficient mice was infected and left undisturbed. Survival was controlled daily.

Evaluation of parasitemia

Bloodstream forms of *T. cruzi* were counted under standardized conditions by direct microscopic observation of 5 µL heparinized blood obtained from the tip of the tail on day 17 p.i. Data are expressed as number of parasites/50 fields.

Corticosterone determination

Plasma samples for hormone determinations were obtained from the tip of the tail under light ether narcosis between 8 and 10 a.m. before injection (time 0), and 7, 14, and 17 days p.i. Corticosterone levels in plasma were determined using a commercially available enzyme-linked immunosorbent assay kit (IBL, Hamburg, Germany).

Determination of TNF-α, IL-1β, and IL-6 levels in plasma

Blood samples were collected after sacrificing the animals 18 days p.i. Plasma was stored frozen at −20 °C until used for the determinations. TNF-α, IL-1β, and IL-6 concentrations were evaluated by ELISA, using commercially available kits (R&D), and the limit of detection was 5.1, 3, and 15.6 pg/mL, respectively. All samples were assayed in duplicate.

Neurotransmitter determination in the spleen, lymph nodes, and heart

Noradrenalin (NA), 5-hydroxytryptamine (serotonin; 5-HT), its precursor tryptophan (Trp), and its main metabolite 5-hydroxyindoleacetic acid (5-HIAA) were determined in the spleen, inguinal lymph nodes, and heart by HPLC with electrochemical detection, as described previously,[5] using the supernatant of tissue samples homogenized in 0.4 M $HClO_4$. Results were expressed as concentration (ng neurotransmitter/g organ), and as the content of the neurotransmitter in the whole organ (ng neurotransmitter/g × organ weight).

Statistical analysis

Results are expressed as means ± SEM. Data were analyzed using one-way analysis of variance (ANOVA) followed by Fisher's test for multiple comparisons or by nonparametric tests (Mann–Whitney *U* test for two samples and Kruskall–Wallis test for *k* samples).

Results

Body weight, parasitemia, and plasma levels of some representative cytokines were evaluated in C57 and RAG-1–deficient mice 18 days after infection with

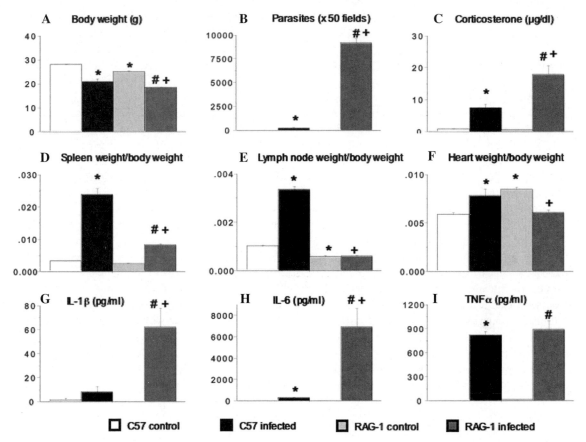

Figure 1. Body, spleen, lymph node, heart weight, parasitemia, corticosterone, and cytokine blood levels following *T. cruzi* infection. C57Bl/6J (C57) and RAG-1–deficient (RAG-1) mice were sacrificed 18 days after injection of 100 trypomastigotes (infected) or the vehicle alone (control). Organ weights are expressed as relative to body weight. Parasitemia and corticosterone plasma levels were evaluated on day 17 p.i. Results are expressed as mean ± SEM of determinations performed in 6–7 mice per group. $^*P < 0.05$ versus C57 control; $^{\#}P < 0.05$ versus RAG-1–deficient control; $^{+}P < 0.05$ versus C57 infected.

T. cruzi or inoculation of the vehicle alone. The body weight was decreased in both types of infected mice at this time (Fig. 1A). The spleen weight increased about fivefold following *T. cruzi* infection of C57 mice and only about threefold in RAG-1–deficient mice, and this difference was maintained when the weight of the organ was expressed relative to the body weight (Fig. 1D). The immune-deficient mice had smaller lymph nodes than C57 mice, and the weight did not increase after infection. In contrast, the weight of the lymph nodes of infected normal mice increased about threefold in infected C57 mice, also when the results were expressed relative to the body weight (Fig. 1E). The heart was chosen because it is a main target of the parasite.[6] Interestingly, while the heart of C57 mice increased

after infection, a reduction was noticed in RAG-1–deficient mice (Fig. 1F).

RAG-1–deficient mice had a parasite load that was 20-fold higher (Fig. 1B), and 10-fold higher IL-1β and IL-6 blood levels than C57 mice; however, although also increased as compared with the noninfected controls, no differences in TNF-α levels were detected (Fig. 1G, H, and I). Corticosterone blood levels on day 17 were also more than twofold higher in infected RAG-1–deficient as compared with infected C57 mice (Fig 1C). Blood levels of this hormone were also determined in these mice before infection (time 0), and 7 and 14 days later (Fig. 2A). No differences in basal levels were detected between noninfected, RAG-1–deficient and C57 mice. Corticosterone levels were already significantly elevated in the immunodeficient mice on day 14 p.i., compared with noninfected controls. All infected RAG-1–deficient mice were dead by day 21 after infection, whereas all infected C57 mice were still alive at this time (Fig. 2B).

NA, 5-HT, and 5-HIAA levels were evaluated on day 18 p.i. in the spleen, inguinal lymph nodes, and heart. NA concentration in the spleen of noninfected RAG-1–deficient mice was significantly higher than in C57 mice (Fig. 3). This difference was no longer significant when values were expressed as total splenic NA content (data not shown), probably due to a nonsignificant tendency of the spleen of RAG-1–deficient mice to be smaller than in normal mice. This way of expressing data of neurotransmitters in the spleen is, however, questionable considering that this organ can accumulate plasma and erythrocytes during the increase in blood flow that parallels immune cell activation.[7] Nevertheless, a statistically significant reduction in splenic NA expressed both as concentration and as total content was observed in both types of mice after infection (Fig. 3). NA concentration in lymph nodes was increased in infected RAG-1–deficient mice when compared with that of noninfected controls and with infected C57 mice. Although NA concentration was decreased in

the heart of *T. cruzi*–infected C57 mice, it was increased in infected RAG-1–deficient mice (Fig. 3).

Serotonin concentration was lower in the spleen of noninfected RAG-1–deficient mice compared with C57 mice (Fig. 4). Following infection, the levels were reduced in C57 but increased in immune-deficient mice. In addition, the basal concentration of the serotonin metabolite 5-HIAA was higher in RAG-1–deficient mice than in C57 as well as the ratio HIAA/5-HT (not shown), indicating an increased serotonin turnover rate. However, there was no difference in this ratio between these two strains after infection. Basal concentrations of tryptophan, the serotonin precursor, were higher in RAG-1–deficient than in C57 mice. After infection, tryptophan concentration increased in both types of mice, but was lower in the immune-deficient than in C57 mice (data not shown).

Basal 5-HT concentration in lymph nodes of RAG-1–deficient mice was about sixfold higher than in C57 mice, and infection resulted in a significant reduction (Fig. 4). No significant differences due to the infection were observed in normal mice, but concentration of 5-HIAA was reduced in both strains of mice. Because changes in both 5-HT and its metabolite were proportional in infected and

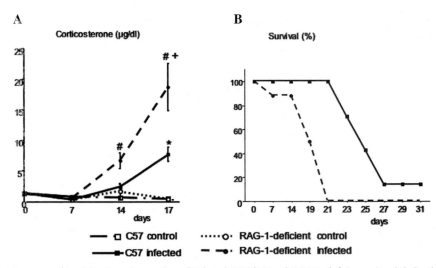

Figure 2. Corticosterone blood levels and mortality of infected C57Bl/6J and RAG-1–deficient mice. (A) Corticosterone levels in plasma were determined in C57Bl/6J and RAG-1–deficient mice (day 0); immediately after, mice were inoculated with vehicle (control) or 100 trypomastigotes (infected). Corticosterone levels in the same mice were also evaluated 7, 14, and 17 days after inoculation. Results are expressed as mean ± SEM. (B) C57 ($n = 7$) and RAG-1–deficient ($n = 8$) were infected with 100 trypomastigotes and left undisturbed. The percentage of mice living at each given time is indicated in the curves. *$P < 0.05$ versus C57 control; #$P < 0.05$ versus RAG-1–deficient control; +$P < 0.05$ versus C57 infected.

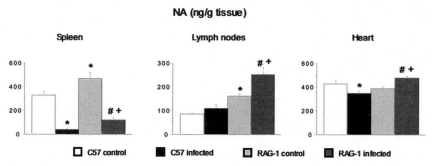

Figure 3. NA concentration in the spleen, inguinal lymph nodes, and heart of C57Bl/6J and RAG-1–deficient mice. C57Bl/6J (C57) and RAG-1–deficient (RAG-1) mice were sacrificed 18 days after injection of 100 trypomastigotes (infected) or the vehicle alone (control). NA concentration was evaluated in the spleen, lymph nodes, and heart. Results are expressed as mean ± SEM of determinations performed in 6–7 mice per group. *$P < 0.05$ versus C57 control; #$P < 0.05$ versus RAG-1–deficient control; +$P < 0.05$ versus C57 infected.

noninfected mice of both types, the 5-HIAA/5-HT ratio was not changed, indicating a balanced production and utilization of serotonin. Tryptophan concentration was lower in the lymph nodes of RAG-1–deficient than in C57 mice, but significantly increased after the infection (data not shown).

Basal serotonin concentration in the heart was lower in RAG-1–deficient than in C57 mice. Infection resulted in decreased 5-HT concentrations in both types of animals, but it was significantly lower in the immune deficient than in the normal mice. A comparable pattern was observed for 5-HIAA concentrations (Fig. 4).

Discussion

RAG-1–deficient mice have been used as a model to explore natural immunity in the absence of adaptive immunity.[8] In the studies reported here, we use this model to explore possible differences in the response of the HPA axis and the levels of NA and 5-HT following infection with *T. cruzi*. The higher parasitemia and earlier mortality of RAG-1–deficient mice inoculated with *T. cruzi* as compared with the wild-type mice attest for the relevance of adaptive immunity in the control of the infection. Such control is mainly achieved by Th1, Th17, and CD8+ T cells that recognize and respond to *T. cruzi*–specific antigens.[4] These defense mechanisms cannot operate in RAG-1–deficient mice because they lack mature B and T lymphocytes. Thus, immune defenses in RAG-1–deficient mice are mainly restricted to innate immunity. This ancient defense mechanism is predominantly activated by a MyD88-dependent mechanism, but there are indications that also TRIF-

dependent innate activation pathways contribute to control *T. cruzi* infection.[4] However, activation of the toll-like receptors TLR2 and TLR9 by PAMP products of *T. cruzi* play a predominant role in host defenses against the parasite. The cross-talk between innate and adaptive immunity is also interrupted in RAG-1–deficient mice.[4]

There is now conclusive evidence that immune responses to infectious agents, for example, during pulmonary tuberculosis,[9] are paralleled by complex neuroendocrine host responses that can affect the operation of immune cells. There are also indications that these neuroendocrine responses are, to a large extent, caused by the immune response itself rather than by the infective agent because they can also be elicited by innocuous, noninfective antigens.[10] Furthermore, several cytokines can affect the functioning of the HPA and other endocrine axes, and of the autonomic nervous system (for review, see Ref. 11). We have previously shown that the HPA axis is strongly activated in normal mice during *T. cruzi* infection.[1] Although the increase in endogenous glucocorticoids induces thymus atrophy, it is to a certain extent beneficial for the host because its interference results in an uncontrolled production of proinflammatory cytokines and in earlier death. We have also detected alterations in the SNS during *T. cruzi* infection of immune competent hosts and found that they are also relevant for the course of the disease (manuscript submitted).

The body weight was comparable in noninfected RAG-1–deficient mice and wild-type C57 mice, indicating that, under protected laboratory conditions, their immunodeficiency is partially

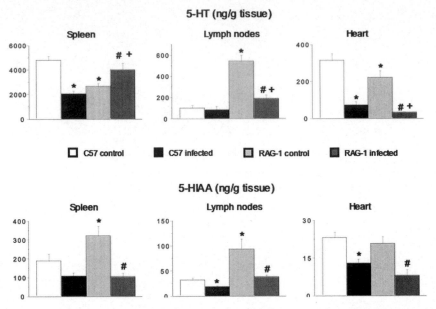

Figure 4. Serotonin and 5-HIAA concentration in the spleen, inguinal lymph nodes, and heart of C57Bl/6J and (RAG-1)–deficient mice. The concentration of serotonin (5-HT) and its main metabolite 5-HIAA was determined in the spleen, lymph nodes, and heart of the same animals from which the results shown in Figure 3 were obtained. Results are expressed as mean ± SEM of determinations performed in 6–7 mice per group. *$P < 0.05$ versus C57 control; #$P < 0.05$ versus RAG-1–deficient control; +$P < 0.05$ versus C57 infected.

compensated. Infected RAG-1–deficient mice had several fold higher parasitemia, and IL-1 and IL-6 blood levels, than infected normal mice. Because it has been reported that serotonin stimulates the production of these cytokines,[12] our results showing that splenic serotonergic activity is enhanced in immune-deficient mice would agree with this finding. However, the same group reported that serotonin inhibits TNF-α production. In our studies, we found that, although also very elevated, TNF-α levels are comparable in both types of mice after infection.

The response of the HPA axis to *T. cruzi* infection occurred earlier and was stronger in RAG-1–deficient than in C57 mice. This effect is most likely caused by exacerbated IL-1β and IL-6 production in infected RAG-1–deficient mice, and indicates that the HPA axis, which, as we have shown operates in normal mice infected with *T. cruzi*,[1] is even overactivated in animals that have no functional T and B lymphocytes.

Basal NA concentrations were higher in the spleen of noninfected RAG-1–deficient than in C57 mice. However, the splenic NA concentration and content were decreased in both after infection. Such de-

crease during prolonged enhanced immune activity is, in general, a reflection of the loss of noradrenergic nerve fibers, as observed, for example, in a model of lymphoproliferative disease,[13] and in target organs of inflammatory process such as arthritic joints.[14] Such loss has also been observed in the spleen of normal mice infected with *T. cruzi* (manuscript submitted). The results reported here indicate that cells other than mature T or B cells, or their products, contribute to this effect.

Even though the NA content in lymph nodes was comparable because the immunodeficient animals have smaller lymph nodes, the concentration of the neurotransmitter was increased in RAG-1–deficient mice, compared with normal mice. This difference is even more marked when animals are infected. This pattern indicates that resident cells, which are not mature T and B lymphocytes, in the lymph nodes of RAG-1–deficient mice are exposed to higher concentrations of NA than in normal mice.

The basal serotonin concentration in the lymph nodes was about fivefold higher in RAG-1–deficient than in normal animals. As in the spleen, the lymph nodes of these immune-deficient animals have also higher basal levels of its metabolite 5-HIAA, indicating that the absence of mature T and B lymphocytes is related to such difference. Following infection, serotonin levels increased in the spleen of the immune deficient mice but decreased in the lymph nodes. However, these low values were still higher when compared with those of infected normal mice. Decreased 5-HIAA concentrations in the spleen and lymph nodes were detected in both types of mice after infection. However, the ratio of 5-HIAA/5-HT only significantly decreased in the spleen of the infected RAG-1–deficient mice, probably indicating reduced utilization of indolamine. Serotonergic nerves were found in the enteric nerve system,[15] pancreas,[16] and iris.[17] However, so far there is no evidence that serotonergic nerve fibres are present in the spleen and lymph nodes, although there are indications that serotonin can be costored with NA and released following sympathetic nerve stimulation.[18] Thus, the most likely source of serotonin in these organs is platelets, which store 5-HT synthesized by enterochromaffin cells in the gut[19] and mast cells.[20] It has been shown that activated T lymphocytes can synthesize serotonin.[21] However, the results reported here indicate that T lymphocyte–derived serotonin seems not to be a main source of this indolamine. Indeed, under basal conditions, RAG-1–deficient mice had even higher concentration of serotonin in the lymph nodes than normal mice, and its concentration was elevated in the spleen even after infection with *T. cruzi*.

Cardiomegaly, a feature of Chagas disease, is attributed to local inflammatory processes linked to TLR2-induced NF-κB activation and local IL-1 production.[6] The increased weight of the heart relative to the body weight observed in C57 mice 17 days after *T. cruzi* inoculation seems to reproduce this situation in the experimental model used. Conversely, a significant decrease in the absolute and relative weight of the heart of RAG-1–deficient mice was observed despite the fact that these animals have more parasites in their blood. Interestingly, it has been reported that anti-β-1 adrenergic receptor antibodies seem to contribute to the pathogenesis of dilated cardiomyopathy and heart failure in Chagas disease.[22–24] These antibodies, which could interfere with anti-inflammatory effects of catecholamines, cannot be produced by RAG-1–deficient mice. The absence of cardiomegaly in RAG-1 mice might therefore be explained by their incapacity to produce anti-β-1 adrenergic receptor antibodies. However, it remains to be explained why, instead of cardiomegaly, the weight of the heart is reduced in *T. cruzi*–infected immune-deficient mice. Such an effect might be related to the much higher production of proinflammatory cytokines in these mice. In fact, it has been shown that IL-1 and TNF-α in particular can induce cardiomyocyte death.[25–29] Furthermore, RAG-1–deficient mice do not produce enough anti-inflammatory cytokines, such as IL-10, that could counteract the proapoptotic effects of IL-1 and TNF-α on cardiac cells.[30,31]

In conclusion, our result suggests that the absence of adaptive immunity, as is the case in RAG-1–deficient mice, results in alterations in some neuroendocrine parameters under basal conditions and in response to *T. cruzi* infection. In our view, this is conceptually relevant not only because it provides further indications that neuroendocrine responses during infections are mediated by immune-derived products, but also because it stresses that the final outcome of these responses is based on interactions between natural and adaptive immunity. Indeed, the role of natural immunity following infection seems to be exacerbated in RAG-1–deficient mice, as reflected by higher levels of IL-1β and IL-6 than in normal mice. As discussed previously, although this was not the case for TNF-α, such increase might be explained by the inhibitory effect of serotonin, which have mixed effects on cytokine production and on natural immunity in general.[18,32] The endogenous production of glucocorticoids is increased earlier and the levels are higher in RAG-1–deficient than in C57 mice, and the results suggest that there are also differences in the activity of the SNS in lymphoid organs, both under basal conditions and following *T. cruzi* infection. At present, we are studying the significance of the different pattern of alterations in the levels of NA and 5-HT in the spleen, lymph nodes, and heart between RAG-1–deficient and C57 mice for the course of the disease. However, the high levels of corticosterone in blood attained in infected RAG-1–deficient mice already indicate that this anti-inflammatory hormone mediates, although not enough, an exacerbated host response to restrict the inflammatory component of the disease caused by the parasite.

Acknowledgments

We thank Pablo Feldman and María Herminia Mogetta for their excellent technical assistance.

Conflicts of interest

The authors declare no conflicts of interest.

References

1. Roggero, E. *et al.* 2006. Endogenous glucocorticoids cause thymus atrophy but are protective during acute Trypanosoma cruzi infection. *J. Endocrinol.* **190:** 495–503.
2. Dutra, W.O. & K.J. Gollob. 2008. Current concepts in immunoregulation and pathology of human Chagas disease. *Curr. Opin. Infect. Dis.* **21:** 287–292.
3. Pellegrini, A. *et al.* 2011. Trypanosoma cruzi antigen immunization induces a higher B cell survival in BALB/c mice, a susceptible strain, compared to C57BL/6 B lymphocytes, a resistant strain to cardiac autoimmunity. *Med. Microbiol. Immunol. (Berl)* **200:** 209–218.
4. Tarleton, R.L. 2007. Immune system recognition of Trypanosoma cruzi. *Curr. Opin. Immunol.* **19:** 430–434.
5. del Rey, A. *et al.* 2003. Sympathetic abnormalities during autoimmune processes: potential relevance of noradrenaline-induced apoptosis. *Ann. N.Y. Acad. Sci.* **992:** 158–167.
6. Petersen, C.A., K.A. Krumholz & B.A. Burleigh. 2005. Toll-like receptor 2 regulates interleukin-1beta-dependent cardiomyocyte hypertrophy triggered by Trypanosoma cruzi. *Infect. Immun.* **73:** 6974–6980.
7. Rogausch, H. *et al.* 2004. The sympathetic control of blood supply is different in the spleen and lymph nodes. *Neuroimmunomodulation* **11:** 58–64.
8. Kim, K.D. *et al.* 2007. Adaptive immune cells temper initial innate responses. *Nat. Med.* **13:** 1248–1252.
9. del Rey, A. *et al.* 2007. Endocrine and cytokine responses in humans with pulmonary tuberculosis. *Brain Behav. Immun.* **21:** 171–179.
10. Besedovsky, H.O. & A. del Rey. 2007. Physiology of psychoneuroimmunology: a personal view. *Brain Behav. Immun.* **21:** 34–44.
11. Besedovsky, H.O. & A., del Rey. 1996. Immune-neuroendocrine interactions: facts and hypotheses. *Endocr. Rev.* **17:** 64–102.
12. Muller, T. *et al.* 2009. 5-hydroxytryptamine modulates migration, cytokine and chemokine release and T-cell priming capacity of dendritic cells in vitro and in vivo. *PLoS One* **4:** e6453.
13. del Rey, A. *et al.* 2006. The role of noradrenergic nerves in the development of the lymphoproliferative disease in Fas-deficient, lpr/lpr mice. *J. Immunol.* **176:** 7079–7086.
14. del Rey, A. *et al.* 2008. Disrupted brain-immune system-joint communication during experimental arthritis. *Arthritis Rheum.* **58:** 3090–3099.
15. Gershon, M.D. 1999. Review article: roles played by 5-hydroxytryptamine in the physiology of the bowel. *Aliment. Pharmacol. Ther.* **13**(Suppl. 2): 15–30.
16. Koevary, S.B., R.C. McEvoy & E.C. Azmitia. 1980. Specific uptake of tritiated serotonin in the adult rat pancreas: evidence for the presence of serotonergic fibers. *Am. J. Anat.* **159:** 361–368.
17. Tobin, A.B., W. Unger & N.N. Osborne. 1988. Evidence for the presence of serotonergic nerves and receptors in the iris-ciliary body complex of the rabbit. *J. Neurosci.* **8:** 3713–3721.
18. Mossner, R. & K.P. Lesch. 1998. Role of serotonin in the immune system and in neuroimmune interactions. *Brain Behav. Immun.* **12:** 249–271.
19. Ahern, G.P. 2011. 5-HT and the immune system. *Curr. Opin. Pharmacol.* **11:** 29–33.
20. Kushnir-Sukhov, N.M. *et al.* 2007. Human mast cells are capable of serotonin synthesis and release. *J. Allergy Clin. Immunol.* **119:** 498–499.
21. Leon-Ponte, M., G.P. Ahern & P.J. O'Connell. 2007. Serotonin provides an accessory signal to enhance T-cell activation by signaling through the 5-HT7 receptor. *Blood* **109:** 3139–3146.
22. Jahns, R., V. Boivin & M.J. Lohse. 2006. Beta 1-adrenergic receptor-directed autoimmunity as a cause of dilated cardiomyopathy in rats. *Int. J. Cardiol.* **112:** 7–14.
23. Labovsky, V. *et al.* 2007. Anti-beta1-adrenergic receptor autoantibodies in patients with chronic Chagas heart disease. *Clin. Exp. Immunol.* **148:** 440–449.
24. Levin, M.J. & J. Hoebeke. 2008. Cross-talk between anti-beta1-adrenoceptor antibodies in dilated cardiomyopathy and Chagas' heart disease. *Autoimmunity* **41:** 429–433.
25. Gordon, J.W., J.A. Shaw & L.A. Kirshenbaum. 2011. Multiple facets of NF-kappaB in the heart: to be or not to NF-kappaB. *Circ. Res.* **108:** 1122–1132.
26. Hedayat, M. *et al.* 2010. Proinflammatory cytokines in heart failure: double-edged swords. *Heart Fail. Rev.* **15:** 543–562.
27. Li, S. *et al.* 2007. Tumor necrosis factor-alpha in mechanic trauma plasma mediates cardiomyocyte apoptosis. *Am. J. Physiol. Heart Circ. Physiol.* **293:** H1847–H1852.
28. Li, Y. *et al.* 2011. Myocardial ischemia activates an injurious innate immune signaling via cardiac heat shock protein 60 and Toll-like receptor 4. *J. Biol. Chem.* **286:** 31308–31319.
29. Suzuki, K. *et al.* 2001. Overexpression of interleukin-1 receptor antagonist provides cardioprotection against ischemia-reperfusion injury associated with reduction in apoptosis. *Circulation* **104:** I308–I303.
30. Pulkki, K.J. 1997. Cytokines and cardiomyocyte death. *Ann. Med.* **29:** 339–343.
31. Reim, D. *et al.* 2009. Role of T cells for cytokine production and outcome in a model of acute septic peritonitis. *Shock* **31:** 245–250.
32. Cloez-Tayarani, I. & J.P. Changeux. 2007. Nicotine and serotonin in immune regulation and inflammatory processes: a perspective. *J. Leukoc. Biol.* **81:** 599–606.

Ann. N.Y. Acad. Sci. ISSN 0077-8923

ANNALS OF THE NEW YORK ACADEMY OF SCIENCES

Issue: *Neuroimmunomodulation in Health and Disease*

Thymic atrophy in acute experimental Chagas disease is associated with an imbalance of stress hormones

Ailin Lepletier,[1] Vinícius de Frias Carvalho,[2] Alexandre Morrot,[3] and Wilson Savino[1]

[1]Laboratory of Thymus Research, Oswaldo Cruz Institute, Oswaldo Cruz Foundation, Rio de Janeiro, Brazil. [2]Laboratory of Inflammation, Oswaldo Cruz Institute, Oswaldo Cruz Foundation, Rio de Janeiro, Brazil. [3]Laboratory of Immunobiology, Institute of Microbiology, Federal University of Rio de Janeiro, Rio de Janeiro, Brazil

Address for correspondence: Wilson Savino, Laboratory on Thymus Research, Department of Immunology, Oswaldo Cruz Institute. Ave. Brasil 4365, Rio de Janeiro, 21045-900 Brazil. savino@fiocruz.br; w_savino@hotmail.com

Disorders in the hypothalamic–pituitary–adrenal axis are associated with the pathogenesis of *Trypanosoma cruzi* infection. During the acute phase of this disease, increased levels of circulating glucocorticoids (GCs) correlate with thymic atrophy. Recently, we demonstrated that this phenomenon is paralleled by a decrease of prolactin (PRL) secretion, another stress hormone that seems to counteract many immunosuppressive effects of GCs. Both GCs and PRL are intrathymically produced and exhibit mutual antagonism through the activation of their respective receptors, GR, and PRLR. Considering that GCs induce apoptosis and inhibit double-positive (DP) thymocyte proliferation and that PRL administration prevents these effects, it seems plausible that a local imbalance of GR–PRLR crosstalk underlies the thymic involution occurring in acute *T. cruzi* infection. In this respect, preserving PRLR signaling seems to be crucial for protecting DP from GC-induced apoptosis.

Keywords: *Trypanosoma cruzi*; thymic atrophy; prolactin; glucocorticoids; CD4$^+$CD8$^+$ double-positive cells

The cross-talk between the neuroendocrine and immune systems is crucial for homeostasis. This interaction occurs because of the sharing of receptors and ligands commonly expressed in both systems. In particular, thymus physiology is highly controlled by hormones, both systemically and locally produced.

Many pathological conditions affect intrathymic T cell development. During acute infection with *Trypanosoma cruzi* (the causative agent of Chagas disease), increased levels of circulating glucocorticoids (GCs) are related to thymus atrophy outcome, characterized by intense immature cortical thymocyte depletion. Despite its deleterious effects in the thymus, high levels of circulating GCs seem to be necessary for protecting the individual from the lethal consequences of an exacerbated proinflammatory response. In this context, we have investigated the role of prolactin (PRL) in preventing GC-induced thymic atrophy during *T. cruzi* infection. Under stressful conditions, PRL is needed to balance the negative effects of GC in order to maintain steady-state homeostasis. To date, it is known that decreased levels of PRL parallel the increase of systemic GC levels. Thus, it seems conceivable that these stress-related hormone imbalances correlate with thymic atrophy during *T. cruzi* infection.

PRL, GCs, and their interactions in the immune system

GCs are steroidal hormones that regulate a range of physiological events, many of them in the immune system, where they exert immunosuppressive and anti-inflammatory actions.[1] In addition to the classical pathway of GC hormonal production that occurs through the activation of the hypothalamus–pituitary–adrenal (HPA) axis, ectopic GC synthesis occurs in the brain, the gastrointestinal tract, and the thymus, where it exerts a paracrine modulation of several functions.[2]

The classical pathway of GC action involves the activation of the glucocorticoid receptor (GR). Being lipophilic hormones, GCs penetrate cells to the cytosol and activate the intracellular GR to translocate into the nucleus where it acts in a multitude of ways to induce or repress the transcription of target genes, including those involved in inflammatory and

doi: 10.1111/j.1749-6632.2012.06601.x

apoptotic responses.[3] The pathways include binding to glucocorticoid response elements (GREs) in regulatory gene regions and protein–protein interactions with other transcription factors.[4] GR inhibition of many proinflammatory response genes occurs through the induction of anti-inflammatory proteins synthesis, as well as through repression of proinflammatory transcription factors, such as nuclear factor-κB (NF-κB) or activator protein-1 (AP-1).[5]

Although being protective, due to the control of an exacerbated proinflammatory response, GC-induced immunossupression can be lethal.[6] Interestingly, PRL works as a stress-adaptation molecule necessary for controlling the negative side effects of GCs and other immune or inflammatory mediators under stressful conditions.[7] Indeed, both PRL and levels of circulating GCs are increased in a variety of immune responses, a phenomenon related to the direct action of increased proinflammatory cytokines in the pituitary gland.[8]

Prolactin is a 23 kDa polypeptide synthesized primarily in the pituitary gland. Similar to GCs, PRL is ectopically synthesized in several tissues, including the immune system, and its proinflammatory effects are related to the activation of a specific prolactin receptor (PRLR).[9,10] A member of the hematopoietin/cytokine receptor superfamily, the activation of this receptor is involved in growth and differentiation of hematopoietic lineages. Once activated, PRLR dimerization induces tyrosine phosphorylation and activation of the cytoplasmic tyrosine kinase, JAK2, followed by phosphorylation of the receptor. One major signaling pathway, characteristic of the long PRLR, involves phosphorylation of cytoplasmic Stats proteins, which themselves dimerize, translocate to nucleus, and bind to specific promoter elements on PRL-responsive genes. The Stats proteins can also physically interact with other transcriptions factors, such as GR.[11]

It seems that the cross-talk between the STAT proteins, especially STAT5, and GR can explain how PRL antagonizes GC effects in target genes. It is known that STAT5 can inhibit GC-mediated activation of GREs through the formation of a complex with the GR that binds to DNA independently of the GRE. This STAT5–GR complex diminishes the GC response of a GRE-containing promoter.

Based on these data, we focus on the role of PRL in preventing GC-induced apoptosis in *T. cruzi* acutely infected mice, raising the hypothesis that the homeostasis of the thymus may be related to the intrathymic balance of GR–PRLR signalings.

PRL antagonizes GC immunosuppressive effects in the thymus

It is largely known that a variety of thymic functions and cellular interactions are under neuroendocrine control.[12] Some of these effects derive from systemic and locally produced hormones. Accordingly, thymic cells express the corresponding hormone receptors. For example, both PRLR and GR are expressed in thymic epithelial cells (TEC) and thymocytes, and the equilibrium of their signaling seems to be crucial for thymus homeostasis.[12]

Pituitary and steroidal hormones oppositely influence the viability and proliferation of thymic cells. While PRL *in vitro* treatment increases thymocyte and TEC proliferation, GC inhibits thymocyte growth.[13] Importantly, the PRL-secreting GH3 pituitary adenoma cells revert the GC-related thymus atrophy in aging rats.[14] Moreover, neonatal thymuses treated with neutralizing anti-PRL antibodies showed a reduction of viable double-positive (DP) cells together with an increase of double-negative (DN) thymocytes,[15] suggesting that PRL regulates thymocyte viability during the DP stage of differentiation and can be involved in rescuing these cells from apoptosis. In fact, PRL protects thymocytes from GC-induced cell death (GICD), both *in vitro*[16] and *in vivo*,[17] likely due to the induction of apoptosis suppressor regulatory genes. Together, these data point to PRL as a stress-adaptation molecule, relevant in maintaining thymus homeostasis in stressful conditions following high GC contents. Accordingly, an imbalance of these hormones may be involved in the thymic involution during acute *T. cruzi* infection.

Neuroendocrine alterations during acute *T. cruzi* infection

Chagas disease is a neglected tropical disease caused by the flagellate protozoan *T. cruzi* and affects millions of people in Latin America. In mice, acute *T. cruzi* infection may lead to encephalitis and promotes dysfunction in the endocrine status, increasing circulating levels of corticosterone, and changing the cytokine profile.[18] This immunoendocrine imbalance is related to distinct pathophysiological aspects of the disease. During the acute phase of *T. cruzi* infection, there is an increased production of the proinflammatory cytokines, interleukin (IL)-1, IL-6, and TNF-α. Despite protecting the organism by stimulating the phagocytic activity of macrophages and destruction of the pathogen, these cytokines can cause fever and even death. Indeed, chagasic myocarditis seems to be associated with an exacerbated Th1 *millieu*. A systemic inflammatory scenario is evident in patients with severe myocarditis compared with healthy subjects.[19] These proinflammatory cytokines activate the HPA axis, causing increased GC release, which contributes to the control of deleterious effects of overproduction of inflammatory mediators.[20]

Additionally, GCs may play a role in the control of autoimmunity in Chagas disease by inhibiting the synthesis of parasite-specific antibodies that are also able to cross-react with host antigens.[21] In this respect, the high levels of circulating GCs are protective during the acute phase of infection, in spite of undesired side effects such as survival of the pathogen and outcome of the disease, which is associated with intense thymic atrophy.[18,22] In this respect, the blockade of GR with the GC antagonist RU486 in *T. cruzi*–acutely infected mice partially reversed the thymic atrophy, but it accelerated death in mice due to increased TNF-α levels.[23] Additionally, otherwise tolerated injections of IL-1 or TNF-α become lethal in adrenalectomized rats, although this is not the case in GC pretreated animals.[24]

Recently, we showed that not only GC but also PRL secretion are altered during Chagas disease.[25] Studying GH3 cells, we found that *T. cruzi* infection decreases PRL secretion, a mechanism that seems to occur through the inhibition of Pit-1, which is thought to be the major cell-specific activator of hormone expression in lactotrophs.

Furthermore, circulating levels of PRL progressively decrease during *T. cruzi* infection (Fig. 1A). This modulation occurs inversely to GC levels, which continuously increase during infection (Fig. 1B). This reverse regulation of PRL and GC secretion seems to be relevant for the GC-induced thymic atrophy seen in the acute phase of *T. cruzi* infection. This stress hormone–related imbalance is related to the presence of the parasite in the adrenal and pituitary glands, which results in disturbances of their endocrine function, as previously reported.[26]

A)

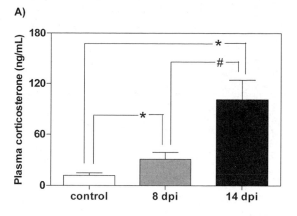

B)

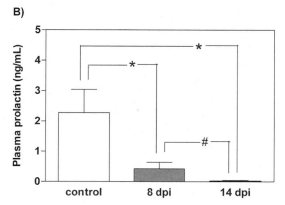

Figure 1. Systemic levels of corticosterone and PRL are inversely modulated during acute *T. cruzi* infection. Corticosterone (A) and PRL (B) were detected by radioimmunoassay and ELISA, respectively, in serum obtained from mice. Infection was carried out using 100 trypomastigote forms of the *T. cruzi* Tulahuén strain. $n = 5$; $^{*}P < 0.05$ related to control; $^{\#}P < 0.05$ related to eight days postinfection (dpi).

Role of GCs in thymic atrophy during acute *T. cruzi* infection

As mentioned previously, increased GC systemic levels are associated with intense thymic atrophy, largely due to the depletion of DP cells, as seen in Figure 2.

The differentiation stages of thymocytes influence their sensitivity to GCs, which do not appear to correlate with the steady state of GR levels, because highly sensitive DP thymocytes express lower levels of GRs than more mature simple-positive (SP) cells, which are relatively GC resistant. Instead, different responses to GCs seem to be related to variations in the GR promoter in thymocyte subsets. Unlike SP subpopulations, DP cells increase GR protein levels

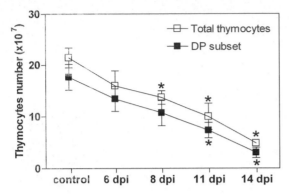

Figure 2. Thymus atrophy during acute *T. cruzi* infection is essentially due to intense DP depletion. The cellularity was obtained from thymuses of infected mice in different days postinfection (dpi). Infection was carried out using 100 trypamastigote forms of the *T. cruzi* Tulahuén strain. Thymocytes were stained with anti-CD4 plus anti-CD8 and phenotypically characterized by flow cytofluorometty. $n = 5$ animals per group; *$P < 0.05$ related to control.

after GC treatment.[27] Indeed, the high levels of circulating GCs during *T. cruzi* infection are associated with DP apoptosis, whereas CD4+ or CD8+ SP cells are not affected (Fig. 3).

The resistance of SP cells also seems to be related to the protective effect of CD28 signaling. The natural ligands of this molecule, CD80/B71 and CD86/B72, are highly expressed in the thymic medulla and stimulate the expression of antiapoptotic proteins of Bcl-2 family in SP thymocytes, which protects these mature cells from GC-induced apoptotic effects. Indeed, SP cells obtained from CD28 knock-out mice are highly susceptible to GICD.[28]

In addition to increasing thymocyte apoptosis in *T. cruzi* acutely infected mice, GC-induced thymus atrophy also seems to be related to the impairment of thymocyte proliferation. The suppression of mitogenic response in thymocytes of infected mice is related to the decrease of IL-2 production due to increased levels of IL-10 and interferon-γ, which are characteristic of Th2 immune responses.[29]

It is important to highlight the possible involvement of intrathymically produced GCs in thymic involution during *T. cruzi* infection. Adrenalectomy of acutely and chronically infected animals only partially prevents thymic atrophy. However, adrenalectomy combined with RU486 treatment prevents not only thymus weight reduction but also the apoptosis-mediated loss of DP thymocytes,[18] re-

flecting the importance of extra-adrenal GCs in this process. In fact, it has been demonstrated that intrathymically produced GCs exert a key role in natural age-dependent thymus involution.[30] We recently found that the fluctuations of intrathymic levels of both corticosterone and PRL in *T. cruzi* acutely infected mice varies independently of their systemic levels (unpublished data). The local consequences for the general process of thymocyte differentiation are under investigation.

Is there an intrathymic GR–PRLR imbalance associated with the thymic atrophy in *T. cruzi* acutely infected animals?

Considering the important role exerted by the mutual antagonism of GR and PRLR signaling in maintaining thymocyte viability in both physiological and stressful situations, it seems plausible to suggest that an imbalance of GR–PRLR cross-talk may be related to the thymic atrophy seen in acutely infected mice. Accordingly, the high susceptibility of DP cells to GICD would be associated with the decrease of PRL-mediated protection.

Besides the deleterious effects in the thymus, the imbalance of these stress hormones seems to be related to the survival and proliferation of the pathogen in the host's target tissues. A recent study demonstrated that PRL administration to infected rats induces T cell proliferation coupled with an

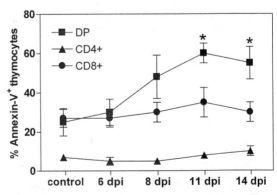

Figure 3. Thymus atrophy during *T. cruzi* infection is related to an increase of apoptosis in DP cells. Thymocytes obtained from infected mice on different days postinfection (dpi) were stained with anti-CD4 and anti-CD8 monoclonal antibodies, as well as annexin-V+, and then phenotypically characterized by cytofluorometry. Infection was carried out using 100 trypamastigote forms of the *T. cruzi* Tulahuén strain. $n = 5$ animals per group; *$P < 0.05$ related to control.

enhancement of macrophage activity and a reduction of blood trypomastigotes during the peak of parasitemia.[31] These results were expected, since PRL triggers proinflammatory immune responses in peripheral immune cells.[32] Unfortunately, the effects of systemic PRL administration on GC synthesis and on the lethal course of *T. cruzi* infection were not explored in this work.

Nothing is known about the role of PRL in thymic atrophy during *T. cruzi* infection. Thymocytes express the long PRLR isoform, which is associated with the activation of JAK/STAT signaling in these cells.[9,10] As mentioned previously, in addition to inducing the expression of antiapoptotic and cell-cycle progression proteins, STAT proteins can abrogate GC-induced immunosuppressive effects through direct interaction with GR. Considering that the expression of GR, in DP thymocytes increases according to its exposition to GCs,[27] it is possible that during acute *T. cruzi* infection high expression of GR, paralleled by decreased PRLR-mediated signaling in these cells, plays a combined role in the generation of thymic atrophy. In this respect, preserving PRLR signaling in DP seems to be crucial for maintaining thymocyte viability, thus preventing thymic atrophy.

Concluding remarks

The immunoendocrine imbalance in human and murine Chagas disease appears to contribute to distinct pathophysiological events, including the thymic atrophy clearly demonstrated in acutely-infected animals. In particular, the data summarized above unravel the possibility that intrathymic and systemic levels of GCs and PRL play mutual roles in thymocyte depletion, with putative effects on the systemic immune response.

Acknowledgments

We thank Dr. Patricia Machado Rodrigues and Silva Martins for kindly discussing our data. This work was partially funded with grants from CNPq, Faperj, and the Oswaldo Cruz Foundation (Brazil).

Conflicts of interest

The authors declare no conflicts of interest.

References

1. Franchimont, D. 2004. Overview of the actions of glucocorticoids on the immune response: a good model to character-ize new pathways of immunosuppression for new treatment strategies. *Ann. N.Y. Acad. Sci.* **1024:** 124–137.
2. Lechner, O. *et al.* 2000. Glucocorticoid production in the murine thymus. *Eur. J. Immunol.* **30:** 337–346.
3. Coutinho, A.E. & K.E. Chapman. 2011. The anti-inflammatory and immunosuppressive effects of glucocorticoids, recent developments and mechanistic insights. *Mol. Cell. Endocrinol.* **335:** 2–13.
4. Mangelsdorf, D.J. *et al.* 1995. The nuclear receptor super-family: the second decade. *Cell* **83:** 835–839.
5. Adcock, I.M. & G. Caramori. 2001. Cross-talk between pro-inflammatory transcription factors and glucocorticoids. *Immunol. Cell Biol.* **79:** 376–384.
6. Oz, H.S. *et al.* 2002. Effects of immunomodulators on acute *Trypanosoma cruzi* infection in mice. *Med. Sci. Monit.* **8:** 208–211.
7. Yu-Lee, L.Y. 2002. Prolactin modulation of immune and inflammatory responses. *Recent Prog. Horm. Res.* **57:** 435–455.
8. McCann, S.M. *et al.* 2000. The mechanism of action of cy-tokines to control the release of hypothalamic and pituitary hormones in infection. *Ann. N.Y. Acad. Sci.* **917:** 4–18.
9. Corbacho, A.M. 2004. Tissue-specific gene expression of prolactin receptor in the acute-phase response induced by lipopolysaccharides. *Am. J. Physiol. Endocrinol. Metab.* **287:** 750–757.
10. Bole-Feysot, C. *et al.* 1998. Prolactin (PRL) and its recep-tor: actions, signal transduction pathways and phenotypes observed in PRL receptor knockout mice. *Endocr. Rev.* **19:** 225–268.
11. Stöcklin, E. *et al.* 1996. Functional interactions between Stat5 and the glucocorticoid receptor. *Nature* **383:** 726–728.
12. Savino, W. & M. Dardenne. 2000. Neuroendocrine control of thymus physiology. *Endocr. Rev.* **21:** 412–443.
13. Timsit, J. *et al.* 1992. Growth hormone and insulin-like growth factor-I stimulate hormonal function and prolifera-tion of thymic epithelial cells. *J. Clin. Endocrinol. Metabol.* **75:** 183–188.
14. Kelley, K.W. *et al.* 1986. GH3 pituitary adenoma cells can reverse thymic aging in rats. *Proc. Natl. Acad .Sci. USA* **83:** 5663–5667.
15. Gaufo, G.O. & M.C. Diamond. 1996. Prolactin in-creases CD4/CD8 cell ratio in thymus-grafted congenitally athymic nude mice. *Proc. Natl. Acad. Sci. USA* **93:** 4165–4169.
16. Biswas, R., T. Roy & U. Chattopadhyay. 2006. Prolactin in-duced reversal of glucocorticoid mediated apoptosis of im-mature cortical thymocytes is abrogated by induction of tumor. *J. Neuroimmunol.* **171:** 120–134.
17. Krishnanan, N. *et al.* 2003. Prolactin suppresses glucocorticoid-induced apoptosis *in vivo*. *Endocrinology* **144:** 2102–2110.
18. Pérez, A.R. *et al.* 2007. Thymus atrophy during *Trypanosoma cruzi* infection is caused by an immuno-endocrine imbal-ance. *Brain Behav. Immun.* **21:** 890–900.
19. Fontanella, G.H. *et al.* 2009. Short treatment with the tu-mour necrosis factor-alpha blocker infliximab diminishes chronic chagasic myocarditis in rats without evidence of *Trypanosoma cruzi* reactivation. *Clin. Exp. Immunol.* **157:** 291–299.

20. Roggero, E. *et al.* 2006. Endogenous glucocorticoids cause thymus atrophy but are protective during acute Trypanosoma cruzi infection. *J. Endocrinol.* **190:** 495–503.

21. Kierszenbaum, F. 2005. Where do we stand on the autoimmunity hypothesis of Chagas disease? *Trends Parasitol.* **21:** 513–516.

22. Savino, W. 2006. The thymus is a common target organ in infectious diseases. *PLoS Pathog.* **6:** 472–483.

23. Roggero, E. *et al.* 2006. Endogenous glucocorticoids cause thymus atrophy but are protective during acute Trypanosoma cruzi infection. *J. Endocrinol.* **190:** 495–503.

24. Bertini, R., M. Bianchi & P. Ghezzi. 1988. Adrenalectomy sensitizes mice to the lethal effects of interleukin 1 and tumor necrosis factor. *J. Exp. Med.* **167:** 1708–1712.

25. Corrêa-de-Santana, E. *et al.* 2009. Modulation of growth hormone and prolactin secretion in *Trypanosoma cruzi*-infected mammosomatotrophic cells. *Neuroimmunomodulation* **16:** 208–212.

26. Corrêa-de-Santana, E., F. Pinto-Mariz & W. Savino. 2006. Immunoneuroendocrine interactions in Chagas disease. *Ann. N.Y. Acad. Sci.* **1088:** 274–283.

27. Purton, J.F. *et al.* 2004. Expression of glucocorticoid expression from 1A promoter correlates to T lymphocyte sensitivity to glucocorticoid-induced cell death. *J. Immunol.* **173:** 3816–3824.

28. van den Brandt, J., D. Wang D & H.M. Reichardt. 2003. Resistance of single-positive thymocytes to glucocorticoid-induced apoptosis is mediated by CD28 signaling. *Mol. Endocrinol.* **18:** 687–695.

29. Leite de Moraes, M.D. *et al.* 1994. Endogenous IL-10 and IFN-gamma production controls thymic cell proliferation in mice acutely infected by *Trypanosoma cruzi*. *Scand. J. Immunol.* **39:** 51–58.

30. Qiao, S. *et al.* 2008. Age-related synthesis of glucocorticoids in thymocytes. *Exp. Cell. Res.* **314:** 3027–3035.

31. Filipin, M. del V. *et al.* 2011. Prolactin: does it exert an up-modulation of the immune response in *Trypanosoma cruzi*-infected rats? *Vet. Parasitol.* **181:** 139–145.

32. Brand, J.M. *et al.* 2004. Prolactin triggers pro-inflammatory immune responses in peripheral immune cells. *Eur. Cytokine Netw.* **15:** 99–104.

Ann. N.Y. Acad. Sci. ISSN 0077-8923

ANNALS OF THE NEW YORK ACADEMY OF SCIENCES
Issue: *Neuroimmunomodulation in Health and Disease*

Endocrine, metabolic, and immunologic components of HIV infection

Guido Norbiato

Department of Endocrinology, L. Sacco University Hospital, Milan, Italy

Address for correspondence: Emeritus Professor Guido Norbiato, viale V. Dandolo 27, 21100 Varese, Italy.
guidonorbiato@libero.it

It is generally accepted that the progression of HIV infection is the consequence of increased HIV virus load and defective CD4$^+$ T cell–mediated immunity. Previous studies have shown that T helper–directed cellular immunity is suppressed in hypercortisolemic HIV patients, while it is activated in cortisol-resistant HIV patients. This is suggestive of a cytokine system intimately linked with cortisol and its receptors. Highly active antiretroviral therapy is an important advance in the treatment of HIV infection, but the suppression of viral replication is not associated with reconstitution of the immune function. This would account for reduced control of inflammation and the activation of 11β-hydroxysteroid dehydrogenase type 1(11β-HSD1) and increases in glucocorticoid and mineralocorticoid production in peripheral tissues. Such hormonal activation may cause insulin resistance and cardiometabolic complications. Therapeutic approaches with 11β-HSD1 inhibitors, aldosterone antagonists, type 1 angiotensin receptor blockers, or renin inhibitors are suggested.

Keywords: immune system; glucocorticoids; cortisol resistance; mineralocorticoids; 11β-hydroxysteroid dehydrogenase type-1, type-2

Introduction

Accumulating evidence indicates that progression of HIV infection is associated with increasing HIV viral load and defects affecting CD4$^+$ T lymphocytes and cell-mediated immunity. Control over progression of HIV infection can be obtained in a small minority of patients in the absence of therapy. In these patients, the infection is associated with a powerful immune response capable of modulating HIV replication.[1] This does not occur in HIV patients treated with highly active antiretroviral therapy (HAART). In fact, although suppression of viral replication is achieved in the majority of individuals treated with HAART, rebound is observed once therapy is interrupted or reduced, demonstrating that a severe immune impairment persists despite therapy.[2] Notwithstanding the great increase in our knowledge of how the immune system is regulated and the interaction between the immune system and the hypothalamic–pituitary–adrenal (HPA) axis, the importance of regulatory links among glucocorticoids, the HPA axis, and the immune system in acquired immune deficiency syndrome (AIDS) is not stressed enough. Cortisol has an important role in the regulation of immune function and metabolism in HIV disease, not only because it contains the exuberance of the infective-inflammatory disease, but also because it enhances or inhibits the immune response, thereby influencing disease susceptibility and progression. Glucocorticoid activity in HIV infection depends not only on glucocorticoid receptor (GR) concentration and binding affinity but also on prereceptor glucocorticoid metabolism and interaction with DNA hormone-responsive elements and other nuclear factors, such as activator protein 1 (AP.1) and nuclear factor κB (NF-κB).[3]

The present review focuses on hormonal and metabolic disorders that may affect HIV-infected individuals, affecting quality of life and potential long-term cardiovascular consequences.

doi: 10.1111/j.1749-6632.2012.06620.x

Such hormonal and metabolic complications may explain the increasing subpopulation of HIV patients seen in endocrine/diabetes clinics.

Hypercortisolism and cortisol resistance in AIDS

It has been noted that during HIV infection, activated immune cells secrete proinflammatory cytokines that stimulate the release of systemic glucocorticoids. The triggering of this regulatory loop during inflammatory processes provides important control over the immune system. However, most frequently in the course of the disease, AIDS patients present several manifestations compatible with an excessive production of cortisol. Further, endogenously released cortisol is an important hormone for controlling the immune system and inflammation. At the forefront of anti-inflammatory and immune-suppressive therapies, glucocorticoids are widely used to treat both acute and chronic inflammation, including rheumatoid arthritis, inflammatory bowel disease, multiple sclerosis, asthma, and psoriasis.[4]

Hypercortisolemia in HIV patients is associated with changes in the immune and T helper (Th)–directed cellular immunity, resulting in low values of interleukin (IL)-2, interferon gamma, and increased values of IL-4 and IgE.[5]

Interestingly, a decreased sensitivity to glucocorticoids has been shown in immune cells of HIV-infected individuals. Competitive binding studies performed on the mononuclear lymphocytes taken from these patients revealed that receptor' affinity for glucocorticoids was reduced, while the number of receptors was increased. Patients with an acquired form of cortisol resistance showed an impressive increase in the Th-directed cellular immune response.[6]

Later on, Vagnucci and Winkelstein[7] showed that more than half of the HIV patients studied have reduced GR-ligand affinity. These findings opened up new directions in the understanding the connection between cortisol and the immune system, in general, and cytokines, in particular. What is made clear by these studies is that the cytokine system in HIV infection is intimately linked with cortisol and its receptors, and that low affinity of cortisol to GR put the AIDS patients in the realm of clinical hypocortisolism, in spite of elevated cortisol plasma levels, similar to those seen in Cushing's disease. Thus,

peripheral cortisol resistance seems, rather than a detrimental complication, an important physiological mechanism aimed to reduce the inhibitory effect of cortisol on the immune system at the expense of its anti-inflammatory action. This interpretation gives a new clinical significance for the response to viral infection and opportunistic infection in the presence of excessive peripheral glucocorticoid activity.

Several distinct molecular mechanisms contribute to decreased cortisol sensitivity in different pathological conditions. Inflammatory diseases, including chronic obstructed pulmonary disease, interstitial pulmonary fibrosis, asthma, cystic fibrosis, and rheumatoid arthritis, appear to be largely glucocorticoid resistant. Dissociated glucocorticoid and selective GR modulators have been developed to reduce the side effects of glucocorticoids, but, so far, it has been difficult to dissociate the anti-inflammatory effects from the adverse effects. Alternative anti-inflammatory treatments are being investigated, as well as are drugs that may reverse the molecular mechanism of glucocorticoid resistance.[8] However, the potential efficacy of such drugs in HIV-induced inflammatory-immune disease is still to be defined.

Effects of HAART

Widespread use of HAART has led to a dramatic and sustained reduction in the morbidity and mortality of HIV infection and has transformed the disease into a chronic condition. HAART treatment usually contains various therapeutic agents that have adverse effects. In particular, administration of HAART is associated with dyslipidemia; lypodystrophy; fat accumulation in the abdomen, the cervical spine (Buffalo hump) and the breast; glucose intolerance; insulin resistance; and diabetes. These patients show a twofold increase in incidence of myocardial infarction; however, this may be only in part attributable to dyslipidemia.[9]

In the beginning it was thought that HAART treatment could reduce HIV viral load and stimulate a quantitatively and qualitatively adequate immune response. However, although suppression of HIV virus was achieved in the majority of HIV-infected individuals, the efficacy of this therapy in reconstituting the immune function was less dramatic.[10] In fact, rebound of HIV replication is observed once therapy is reduced or interrupted. Thus, the absence of virus eradication impedes the immune stimulation of HIV-specific T cell clones and reduces the efficacy of HAART. Acute HIV infection is, therefore, transformed into a chronic, infectious inflammatory disease associated with increased levels of polyunsaturated free fatty acids; lipogengic proinflammatory cytokines IL-1, IFN-α, IFN-γ, and TNF-α; and serious metabolic complications. TNF-α has a particularly important role in the regulation of a cascade of events in a number of inflammatory diseases, exemplified by the induction of cytokines such as IL-1β and IL-6, and of acute-phase proteins such as reactive proteins. TNF-α is also an important trigger of the positive and negative feedback loop, which governs chronicity and pathogenic outcomes of inflammation. Many of these functions are inhibited by TNF-α–specific antagonist.[11] TNF-α and IFN-α are known to stimulate cortisol, which in turn shifts the cytokine profile from a Th1 type (proinflammatory) to a Th2 type (anti-inflammatory). These side effects occur in HIV-infected patients treated with HAART despite normal circulating levels of cortisol and HPA axis function. Thus, evidence suggests that once cytokines have stimulated glucocorticoid release, to keep the inflammatory response in check, it is important to regulate the resulting activity of the glucocorticoids, as either too much or too little activity from steroid hormones may result in pathological consequences.

Prereceptor regulation of glucocorticoids by 11β-hydroxysteroid dehydrogenase type 1

The type 1 isozyme 11β-HSD (11β-HSD1) predominantly acts as a reductase, transforming cortisone into cortisol. This enzyme has been found in a number of tissues, including liver and adipose tissues. Overexpression of 11β-HSD1 may be caused by proinflammatory cytokines, particularly IL-1 and TNF-α, which are elevated in HIV patients treated with HAART.[12] This enzyme is also activated during hyperglycemia. Therefore, the clinical consequences are important, as it implicates a link between glucose disposal and local production of cortisol. The effect is mediated by an increase in cytosolic glucose-6-phosphate, a substrate for hexose-6-phosphate dehydrogenase, which in turn drives 11β-HSD1 activity and local glucocorticoid generation.[13] 11β-HSD1 activation leads to dyslipidemia,

hypertriglyceridemia, adiposity, fat redistribution, insulin resistance, hypertension, and atherosclerosis. Conversely, inhibition or deficiency of 11β-HSD1 prevents progression of atherosclerosis and cardiovascular complications.[12]

Prereceptor regulation of mineralocorticoids: 11β-hydroxysteroid dehydrogenase type 2

11β-Hydroxysteroid dehydrogenase type 2 (11β-HSD2) protects the mineralocorticoid receptors (MR) from cortisol excess, often caused by the activation of 11β-HSD1 by converting cortisol into cortisone. Compromised 11β-HSD2 activity due to genetic mutation, the presence of inhibitors, or saturating cortisol concentrations leads to cortisol-induced activation of MR and transduction of the MR to the nucleus. By contrast, cortisone blocks aldosterone-induced MR activation. Thus 11β-HSD2, besides inactivating cortisol by its conversion to cortisone, functionally interacts with MR and directly regulates the magnitude of aldosterone production.[14] 11β-HSD2 enzyme is mainly expressed in the kidney and other mineralocorticoid-sensitive tissues such as vascular endothelium and heart. 11β-HSD2 enzyme has been also detected in rheumatoid arthritis and peripheral mononuclear cells. Deficiency of 11β-HSD2 results in cortisol-mediated sodium retention and hypertension, suggesting that the physiological regulation of 11β-HSD2 in mineralocorticoid target tissues may be important in modulating sodium homeostasis and blood pressure.[15]

Increased 11β-HSD2 activity may be associated with decreased insulin sensitivity. In this case, the kidney intensifies its supply of cortisone substrate for extrarenal 11β-HSD1, which may fuel visceral adiposity and insulin resistance.[16] Accumulating evidence indicates that cardiovascular, renal abnormalities, and insulin resistance, associated with increased 11β-HSD1 and/or 11β-HSD2 activities, are mediated in part by genomic and not genomic aldosterone signaling through MR activation. In cardiometabolic syndrome, there is an increased production of glucocorticoids, which directly activates MR signaling in cardiovascular, adipose, skeletal muscle, neuronal, and liver tissues.

Recently it has been shown that MR blockade improves the release of pancreatic insulin as well as reduces the progression of cardiometabolic and kidney disease, insulin-mediated glucose use, and endothelium-dependent vasorelaxation. These studies clearly suggest that an excess of aldosterone exerts detrimental metabolic effects that contribute to the development of resistant hypertension as well as cardiovascular and chronic kidney disease.[17]

The renin–angiotensin–aldosterone system in AIDS

The renin–angiotensin–aldosterone system may be activated in HIV-infected patients with hypertension and metabolic–vascular complications.[18] It has also been shown that type-1 angiotensin receptor blockers prevent the activation of angiotensin 2. Further, the angiotensin-receptor blocker thermisartan is much more active than other antihypertensive agents in reducing hypertension and metabolic disturbance in AIDS patients.[19] Angiotensin-converting enzyme inhibitors increase 11β-HSD2 activity, which may explain increases in renal sodium excretion independent of circulating angiotensin-2 concentration.[15]

Discussion

The fundamental concept is that strong and continuous immune activation, associated with low viremia, is necessary for the control of HIV infection. This does not occur in HIV patients treated with HAART, who do not maintain adequate immune activity during treatment, thus explaining why patients have a rebound of HIV replication once the therapy is interrupted, even after a prolonged period of suppression. Persistence of virus replication and inflammation triggers the shuttle of cortisone/cortisol, which amplifies intracellular levels of glucocorticoids. The consequence is that 11β-HSD1 oxoreductase remains high in macrophages, liver, adipose, and muscle cells until the inflammation is resolved.[12] An excessive production of cortisol promotes insulin resistance and metabolic, hormonal, and vascular complications. Activation of 11β-HSD1 also leads to MR activation since, in the presence of scant production of cortisone, cortisol acts as a mineralocorticoid. In addition to inactivating cortisol, 11β-HSD2 interacts with the MR and directly regulates the magnitude of aldosterone production.

Aldosterone, through its nongenomic action, promotes tissue proliferation, fibrosis, and inflammation that have been linked to cardiovascular and cardiometabolic complications. Hyperaldosteronism plays a key role in impairing insulin activity and glucose homeostasis, contributing to the development of endothelial dysfunction, atherosclerosis, kidney disease, and diabetes. MR blockage restores insulin sensitivity, counterbalances the deleterious effects of aldosterone, and presents an alternative to improve blockage of the renin–angiotensin–aldosterone system.[20] An interesting observation is that protease inhibitors that are part of HAART treatment activate the adipocyte renin–angiotensin system, contributing to insulin resistance and vascular complications.[18] Also, type 1 angiotensin 2 receptor blockers are active in treating hypertension and metabolic complicances in HIV-infected patients. Finally, the prorenin receptor–dependent intracellular signal, another member of the system regulated by angiotensin, may be detected in many tissues, including adipose tissues and kidney. Prorenin receptors are preregulated in the diabetic kidney and have been implicated in the glucose-induced overproduction of profibrotic molecules by mesangials cells, which may be involved in vascular complications.[21]

Conclusion

Compelling evidence has been amassed indicating that in patients with HIV infection, the cytokine system is intimately linked with cortisol and its receptors. As immunological responses appear limited during HAART treatment, chronic inflammatory reactions persist, which are counteracted by activation of the cortisone/cortisol shuttle. The excessive production of cortisol due to the activation of the 11β-HSD1 enzyme inhibits the immune response. Interactions between 11β-HSD1 and 11β-HSD2 enzymes lead to increases in both glucocorticoid and mineralocorticoid hormone production, which may play a key role in impairing insulin activity, glucose homeostasis, and contribute to the development of metabolic and vascular complications. Because current treatment strategies have failed to provide effective solutions, these data suggest a rationally based therapy employing specific 11β-HSD1 inhibitors that are expected to reduce peripheral cortisol production and, by analogy with cortisol-resistant AIDS patients, to stimulate immune function. The potential efficacy of the TNF-α and IL-1 antagonists, in a variety of immune-mediated inflammatory diseases, have garnered attention for their possible use in HAART-treated patients.

Finally, aldosterone inhibitors, which are possibly associated with type-1 angiotensin blockers or renin inhibitors, may be effective in preventing and treating insulin resistance, atherosclerosis, and cardiometabolic complications. It is important to recall that, worldwide, a great number of HIV-infected people cannot receive lifetime treatment with antiviral drugs because of the expense. Thus, new treatment strategies are needed to keep HIV virus replication and inflammation under immune system control. We are currently investigating this promising area of research.

Conflicts of interest

The author declares no conflicts of interest.

References

1. Rosemberg, E.S. *et al.* 1997. Vigorous HIV-1 specific cd4+T-cell response associated with control of viremia. *Science* **278:** 147–150.
2. Clerici, M. *et al.* 2000. Different immunologic profiles characterize HIV infection in highly active retroviral therapy–treated and antiretroviral-naïve patients with undetectable viraemia. The Master Group. *Aids* **14:** 109–116.
3. Bamberger, C.M. *et al.* 1996. Molecular determinant of glucocorticoid receptors function and tissue sensitivity to glucocorticoids. *Endocr. Rev.* **17:** 245–261.
4. Coutinho, A.E. & K.E. Chapman. 2011. The anti-inflammatory and immune-responsive effects of glucocorticoid. Recent development and mechanicistic insight. *Mol. cell Endocrinol.* **335:** 2–13.
5. Norbiato, G. 1994. *The Endocrinology and Metabolism of HIV Infection.* Baillier Tindall. Oxford.
6. Norbiato, G. *et al.* 1992. Cortisol resistance in acquired immune deficiency sindrome. *J. Clin. Endocrinol. Metab.* **74:** 608–613.
7. Vagnucci, A.H. & A. Winkelstein. 1993. Circadian rhythm of lymphocytes and their glucocorticoids receptors in HIV-infected homosexual man. *J. Acq. Imm. Defic. Synd.* **6:** 1238–1247.
8. Barnes, P.J. 2011. Glucocorticoids: current and future directions. *Br. J. Pharmacol.* **163:** 29–43.
9. The DAD study group. 2007. Class of antiretroviral drugs and the risk of myocardial infarction. *N. Engl. J. Med.* **356:** 17.
10. Angel, J.E. *et al.* 1998. Improvement in cell mediated immune function during potent anti-human immune-deficiency virus therapy with ritonavir plus sequinavir. *J. Infect. Dis.* **177:** 808–904.
11. Wallis, R.S. 2011. Byologics and infections: lessons from tumor necrosis factor blocking agents. *Infect. Dis. Clin. North Am.* **25:** 895–910.
12. Chapman, K.E. *et al.* 2006. Local amplification of glucocorticoids by 11β-hydroxysteroids dehydrogenase type 1 and its role in the inflammatory response. *Ann. N.Y. Acad. Sci.* **1088:** 265–273.
13. Basu, R. *et al.* 2006. Effect of nutrient ingestion on total body and splanchnic cortisol production in humans. *Diabetes* **55:** 667–674.
14. Odermatt, A. *et al.* 2001. The intracellular localization of the mineralocorticoid receptors is regulated by 11beta-hydroxysteroid dehydrogenase type 2. *J. Biol. Chem.* **276:** 28484–28492.
15. Richetts, M.L. & P.M. Stewart. 1999. Regulation of 11β-hydroxysteroid dehydrogenase type-2 by diuretics and the renin-angiotensin-aldosterone axis. *Clin. Sci.* **96:** 669–675.
16. Müssig, K. *et al.* 2008. 11β-hydroxysteroid dehydrogenase type-2 activity is elevated in obesity and negatively associated with insulin sensitivity. *Obesity* **16:** 1256–1260.
17. Whaley Cornell, A. *et al.* 2010. Aldosterone role of the cardiometabolic syndrome and resistant hypertension. *Prog. Cardiovascular Dis.* **52:** 401–409.
18. Bakkara, F. *et al.* 2010. HIV protease inhibitors activate adipocytes renin-angiotensin system. *Antivir. Ther.* **15:** 363–375.
19. Vecchiet, J. *et al.* 2011. Antihypertensive and metabolic effect of thermisartan in hypertensive HIV positive patients. *Antivir. Ther.* **16:** 639–645.
20. Lastra-Gonzales, S.G. *et al.* 2008. The role of aldosterone in cardiovascular disease in people with diabetes and hypertension: an update. *Curr. Diab. Rep.* **8:** 203–207.
21. Pereira, L.G. *et al.* 2012. (Pro)renin receptors: another member of the system controlled by angiotensin II? *J. Renin–Angiotensin–Aldosterone Syst.* **13:** 1–10.

Ann. N.Y. Acad. Sci. ISSN 0077-8923

ANNALS OF THE NEW YORK ACADEMY OF SCIENCES

Issue: *Neuroimmunomodulation in Health and Disease*

Autoimmune diseases and infections as risk factors for schizophrenia

Michael E. Benros,[1,2] Preben B. Mortensen,[1] and William W. Eaton[3]

[1]National Center for Register-based Research, Aarhus University, Aarhus, Denmark. [2]Mental Health Center Copenhagen, Copenhagen University, Copenhagen, Denmark. [3]Department of Mental Health, Johns Hopkins Bloomberg School of Public Health, Baltimore, Maryland

Address for correspondence: Michael Eriksen Benros, National Centre for Register-based Research, Aarhus University, Taasingegade 1, 8000 Aarhus C, Denmark. Benros@ncrr.dk

Immunological hypotheses have become increasingly prominent when studying the etiology of schizophrenia. Autoimmune diseases, and especially the number of infections requiring hospitalization, have been identified as significant risk factors for schizophrenia in a dose–response relationship, which seem compatible with an immunological hypothesis for subgroups of patients with schizophrenia. Inflammation and infections may affect the brain through many different pathways that are not necessarily mutually exclusive and can possibly increase the risk of schizophrenia in vulnerable individuals. However, the findings could also be an epiphenomenon and not causal, due to, for instance, common genetic vulnerability, which could be supported by the observations of an increased prevalence of autoimmune diseases and infections in parents of patients with schizophrenia. Nevertheless, autoimmune diseases and infections should be considered in the treatment of individuals with schizophrenia symptoms, and further research is needed of the immune system's possible contributing pathogenic factors in the etiology of schizophrenia.

Keywords: schizophrenia; autoimmune disease; infection; inflammation; epidemiology

Introduction

The etiology of schizophrenia is complex and multifactorial, with a psychiatric family history being a main contributing risk factor that is, however, unlikely to be a sufficient cause.[1] Immunological hypotheses have recently become increasingly prominent in psychiatric research, and schizophrenia in particular.[2] Many diverse immune alterations have been observed in patients with schizophrenia,[2–4] and schizophrenia has additionally been associated with genetic markers related to autoimmune diseases and infections.[5,6] Both autoimmune diseases and infections have been suggested to be causally linked with schizophrenia because of inflammatory mechanisms, which possibly affect the brain through many different pathways that are not necessarily mutually exclusive.[7–11] Infections and toxins can induce molecular mimicry with reactivity against the body's own tissue and are believed to be central in the development of autoimmune diseases. In some autoimmune diseases

there is a high prevalence of neuropsychiatric symptoms, possibly induced by brain-reactive antibodies.[12–14] Stress, infections, and inflammation are among the range of possible insults that can compromise the blood–brain barrier (BBB) and might permit immune components to affect the central nervous system (CNS).[15] Abnormalities of the BBB have been indicated in studies of patients with schizophrenia,[16] and CNS inflammation together with blood cerebrospinal fluid (CSF) barrier dysfunction have also been observed.[17] Furthermore, studies have indicated increased autoantibody reactivity and elevated autoantibody levels in patients with schizophrenia.[18,19] Peripheral inflammation or infections can additionally interact with the neuroendocrine function and receptors that are thought to be involved in schizophrenia through many other pathways. In summary, vulnerable individuals might be at risk of developing neuropsychiatric symptoms like schizophrenia as a consequence of inflammation affecting the brain. The aim of this paper is to review associations between autoimmune

doi: 10.1111/j.1749-6632.2012.06638.x

diseases and infections as risk factors for schizophrenia, and the possible etiological mechanisms for the raised risk.

Associations between autoimmune diseases and the risk of developing schizophrenia spectrum disorders

There is more than a half century of research on the association of autoimmune diseases and schizophrenia. As early as the 1950s, investigators were puzzled by the apparent protective effect of schizophrenia for rheumatoid arthritis,[20,21] and more than a dozen studies also show the same negative association.[22] In the 1950s and 1960s, clinicians noticed what seemed to be an unusually high occurrence of celiac disease in persons with schizophrenia,[23,24] and this observation was followed by a number of epidemiologic studies and clinical trials, all of which were consistent with the idea that in-

gestion of gluten causes schizophrenia in a subpopulation with some sort of vulnerability.[25] Recently, it has been found that antibodies to the self-antigen tissue transglutaminase, indicative of celiac disease, are found in about five times as many persons with schizophrenia as in the general population.[26–29] Antibodies to gliadin, indicating sensitivity to wheat not necessarily associated with autoimmune disease, are found in much higher proportions in persons with schizophrenia than in the general population.[27,29,30] A wide range of neurological complications is associated with antibodies to gliadin, even in the absence of autoimmune disease.[15,31,32]

In the last decade, a wider range of autoimmune diseases has been implicated in population-based prospective studies as raising risk for schizophrenia. The relative risk of schizophrenia for an individual with a history of autoimmune disease, including themselves or relatives, was about 45% in an analysis

Table 1. Relative risks of schizophrenia for those with and without a familial or individual history of autoimmune disease

	Risk of schizophrenia with a family history of autoimmune diseases ($n = 16,722$)[a]		Risk of schizophrenia with a personal history of autoimmune diseases ($n = 20,317$)		
				Case concurrent	Case delayed
	Number of cases	Relative risk[b]	Number of cases[c]	Relative risk[c]	
Any of 30 autoimmune diseases	1,205	**1.2**	228/238	**1.4**	**1.3**
Thyrotoxicosis	142	1.1	28/18	**1.9**	**2.1**
Type 1 diabetes	475	**1.3**	59/49	**1.7**	1.0
Multiple sclerosis	141	**1.3**	11/8	1.0	0.8
Guillain-Barre	45	1.1	4/10	–	**2.2**
Iridocyclitis	89	**1.4**	17/10	**1.8**	1.3
Crohn's disease	81	0.9	11/28	0.7	**1.7**
Autoimmune hepatitis	45	**1.7**	12/15	**5.4**	**5.6**
Celiac disease	22	1.2	4/5	–	1.8
Psoriasis vulgaris	158	**1.2**	28/23	**2.0**	1.4
Rheumatoid arthritis	273	1.1	12/12	0.8	0.8
Dermatopolymyositis	20	2.1	–	–	–
Sjogren's syndrome	33	**1.5**	6/1	**3.8**	–

NOTE: Ref. 33.
[a]Family history could not be ascertained for 3,595 cases.
[b]Relative risks are adjusted for age and interaction with sex, calendar year, and ages of the mother and father at the birth of the offspring. Boldface indicates that the 95% confidence interval did not include 1.0. Relative risks were not estimated when there were less than five exposed cases.
[c]Number of cases with concurrent onset/number of cases with delayed onset; total number with prior autoimmune disease is the sum of the two numbers.

of 7,704 cases.[11] In an analysis of a longer time series of 20,317 cases (including the 7,704), there were about a dozen autoimmune diseases that predicted onset of schizophrenia.[33] Table 1 reflects the statistically significant positive predictions among the 30 autoimmune diseases studied, with celiac disease and rheumatoid arthritis included even though the results were not significant for these diseases in this analysis. The raised risk occurs with autoimmune diseases that present in the endocrinologic, neurologic, vascular, dermatologic, and connective tissue systems. In the table, the raised risk in the cases of schizophrenia who have a personal history of autoimmune diseases is divided between concurrent (diagnosis of schizophrenia within 4 years after autoimmune disease) and delayed onset (diagnosis of schizophrenia 5 years or more after autoimmune disease), to assess the possibility that the raised risk might be due to ascertainment bias, which the pattern does not suggest. These results were replicated in the broader category of nonaffective psychosis ($n = 39,067$).[33] The association of a range of autoimmune diseases with schizophrenia, including a positive association with celiac disease and a negative association with rheumatoid arthritis, has been replicated in a recent analysis of a national sample from Taiwan.[34]

Since the early 1960s, a variety of autoantibodies with cross-reactivity against brain antigens have been described in the sera of patients with schizophrenia.[35–37] Several groups have identified diverse brain-reactive antibodies in patients with schizophrenia, including antibodies against neurotransmitters, but consistency in the findings between research groups has not been high,[7,38] and the correlation with disease activity ambiguous.[38] In a recent prospective Danish population-based study, the highest risk of schizophrenia was found in the combined group of autoimmune diseases with suspected presence of brain-reactive antibodies (Table 2).[39] Hospital contact with autoimmune diseases had occurred in 2.4% of patients with schizophrenia before the diagnosis in a Danish setting,[39] and because many autoimmune diseases have clinical onset later than schizophrenia, the number possibly underestimates the effect due to undiagnosed illness. An excess prevalence of diverse autoantibodies has been detected in the sera[18,19,40] and in the CSF[17] of patients with schizophrenia who have not previously been diagnosed with an autoimmune disease.

Furthermore, newly discovered brain-reactive antibodies have been identified in the sera of patients with schizophrenia,[41] and a large amount of previously unknown autoantibodies has been detected.[42]

Associations between infections and the risk of developing schizophrenia spectrum disorders

The possibility that bacterial infections have a causal relationship to psychoses was reported as early as 1896,[43] and virus infections have been suspected since the 1918 influenza pandemic preceded multiple reports of postinfluenza psychoses and schizophrenia-like symptoms.[44] Interest in infectious theories of psychiatric disorders waned in the 1930s due to a lack of relevant treatment methods the growing prominence of Freudian theories, but during the last few decades, research on the relationship between infections and schizophrenia has reemerged.[9] Epidemiological studies have indicated a dose–response relationship between urbanicity during upbringing and the risk of schizophrenia,[45] which could be related to an increased probability of acquiring infections.[9] Most studies have focused on infections during maternal pregnancy and how these infections might affect the fetus and contribute to the future development of schizophrenia in the child.[9,46] Studies have suggested that the effect of infections during maternal pregnancy could be mediated through inflammatory processes rather than through the infection itself because of the failure to consistently find any one specific infectious agent linked to schizophrenia. However, a recent Danish epidemiological study of parental infections showed no significant difference in increased risk of schizophrenia if the infections occurred during pregnancy or outside the pregnancy period, and the risk of schizophrenia was similarly increased when comparing infections in the father or mother.[47] The study might indicate a shared genetic vulnerability to acquire severe infections with hospital contacts in parents of persons with schizophrenia.[47]

An increased risk of schizophrenia has been associated with many different infectious agents, but studies have produced conflicting results. A recent meta-analysis found significant associations between schizophrenia and *Toxoplasma gondii*, human herpesvirus 2, Borna disease virus, human endogenous retrovirus W, *Chlamydophila psittaci*, and *Chlamydophila pneumonia*.[48]

T. gondii infection has been associated with schizophrenia,[9,10,49] and a recent study indicated a dose–response relationship correlating the serum titer of *Toxoplasma* with the subsequent risk of developing schizophrenia.[50] An increased risk of schizophrenia has also been associated with herpes simplex virus infection, both detected in serum antibodies[51,52] and CSF antibodies.[53] Cytomegalovirus (CMV) antibodies have been shown to be higher in the serum of patients with schizophrenia[44,54] and in the CSF,[55] although contradictory associations have been reported between CMV and schizophrenia, with stronger associations in newly diagnosed and untreated patients.[54] To date, no neuropathological evidence of CMV in the brains of patients with schizophrenia has been found.[44] Retroviral antigens and products have been identified in some patients with schizophrenia,[56,57] as has increased evidence in serum of Bornia virus.[58] An increased prevalence of *Chlamydophila* infection has also been observed in patients with schizophrenia, especially when linked to genetic markers of the immune system.[59] In addition, postmortem studies have found increased prevalence of *Chlamydophila*

Table 2. Incidence rate ratios (IRRs) of schizophrenia spectrum diagnosis[a] in persons with a hospital contact of autoimmune diseases and infections, Denmark, 1977–2006

Autoimmune disease	Schizophrenia spectrum disorders in persons without infections			Schizophrenia spectrum disorder in persons with hospital contact with infections		
	IRR[b]	95% CI	Cases	IRR[b]	95% CI	Cases
Persons without autoimmune disease (reference)	1.00	(Reference)	29,372	**1.60**	1.56–1.64	8,777
Any autoimmune disease	**1.29**	1.18–1.41	483[c]	**2.25**	2.04–2.46	444[c]
Autoimmune diseases with suspected presence of brain-reactive antibodies	**1.48**	1.31–1.68	244	**2.56**	2.25–2.89	243
Autoimmune hepatitis	**2.75**	1.38–4.83	10	**8.91**	6.50–11.84	43
Autoimmune thyroiditis	–	–	3	**4.57**	2.09–8.51	8
Celiac disease	**2.11**	1.09–3.61	11	**2.47**	1.13–4.61	8
Guillain-Barre syndrome	1.22	0.58–2.19	9	**2.84**	1.52–4.76	12
Multiple sclerosis	**1.44**	1.03–1.94	39	**2.10**	1.37–3.06	24
Sjogren's syndrome	2.07	0.82–4.20	6	–	–	4
Systemic lupus erythematosis	1.84	0.92–3.23	10	**2.11**	1.06–3.70	10
Thyrotoxicosis (Graves disease)	**1.94**	1.47–2.49	56	**2.47**	1.68–3.49	29
Type 1 diabetes	**1.27**	1.04–1.53	104	**2.04**	1.68–2.44	109
Other autoimmune diseases[d]	**1.19**	1.05–1.34	256	**1.95**	1.70–2.23	212

NOTE: From Ref. 39.
[a]Persons with schizophrenia or schizophrenia-like psychoses (and including schizotypal personality disorder) were included (ICD-8: 295, 296.89, 297, 298.39, 301.83; ICD-10: F20–F29).
[b]Boldface indicates that the 95% confidence interval did not include 1.0. Relative risks were not estimated when there were less than five exposed cases. Relative risks are adjusted for sex, age, and calendar period.
[c]Cases do not add up as one can have multiple autoimmune diseases.
[d]Alopecia areata, ankylosing spondylitis, autoimmune hemolytic anemia, Crohn's disease, dermatopolymyositis, idiopathic thrombocytopenic purpura, iridocyclitis, juvenile arthritis, myasthenia gravis, pernicious anemia, primary adrenocortical insufficiency, primary biliary cirrhosis, pemphigus. pemphigoid, polymyalgia rheumatica, psoriasis, vulgaris, scleroderma, seropositive rheumatoid arthritis, ulcerative colitis, vitiligo, Wegener's granulomatosis.

DNA in brains of patients with schizophrenia compared with controls.[60] Most studies have found an increased risk of schizophrenia in individuals with childhood CNS infections.[8,61,62] However, one study did not find an association when using a control group of children hospitalized for gastroenteritis infection,[63] but one could argue that a gastroenteritis infection might also affect the brain; and as described in Table 3, the risk of schizophrenia has been reported to increase after hospital contact with a CNS infection, and even more after most other types of peripheral infections.[39]

Because no clear relationship between one specific infectious agent and schizophrenia has been identified, it has been suggested that the causal factor is the resulting immune response influencing the CNS.[64] The first large-scale study on infections in general and the subsequent risk of schizophrenia is a recent Danish prospective nationwide cohort study, indicating that hospital contact with an infection significantly increased the risk of schizophrenia by 60% (Table 2). Each hospital contact with infection increased the risk of schizophrenia significantly in a dose–response relationship (Fig. 1).[39] Hospital contacts with infections occurred in nearly 24% of patients with schizophrenia previous to the diagnosis. The population-attributable risk (the fraction of the total number of schizophrenia cases that would not have occurred if the effect of hospital contacts with infections could be eliminated) is estimated at 9%. This is probably an underestimation of the effect because less severe infections treated only by primary care practitioners, or not treated at all, do not enter the Danish registers. But one could argue that the results do not generalize to less severe infections because severe infections might have a higher proportion of sepsis or a more extensive inflammatory response, and therefore are more likely to affect the brain. The risk of schizophrenia increased with the temporal proximity of the infection but still

Table 3. Incidence rate ratios (IRR) of schizophrenia spectrum diagnosis[a] depending on the site of infection in persons with and without autoimmune diseases, Denmark, 1977–2006

Site of first infection	Only infection (no autoimmune disease)			Autoimmune disease		
	IRR[b]	95% CI	Cases	IRR[b]	95% CI	Cases
All infections	**1.60**	1.56–1.64	8,777	**2.25**	2.04–2.46	444
Sepsis infections	**1.95**	1.47–2.51	55	**4.98**	2.49–8.73	10
Hepatitis infections	**4.89**	4.26–5.58	212	**8.89**	6.03–12.53	29
Gastrointestinal infections	**1.32**	1.26–1.39	1,847	**1.82**	1.46–2.24	83
Skin infection	**1.71**	1.62–1.80	1,427	**2.14**	1.69–2.66	74
Pregnancy-related infection	1.14	0.98–1.31	185	1.22	0.48–2.47	6
Respiratory infections	**1.53**	1.46–1.61	1,885	**2.25**	1.79–2.79	77
Urogenital infections	**1.90**	1.79–2.01	1,200	**2.70**	2.10–3.41	66
CNS infections	**1.28**	1.09–1.50	148	**2.62**	1.31–4.60	10
Other types of infections	**1.70**	1.62–1.78	1,818	**1.99**	1.60–2.43	89
Persons without a hospital contact with infection (reference)	1.00	(Reference)	29,372	**1.30**	1.18–1.42	483

[a]Persons with schizophrenia or schizophrenia-like psychoses (and including schizotypal personality disorder) were included (ICD-8: 295, 296.89, 297, 298.39, 301.83; ICD-10: F20–F29).
[b]Boldface indicates that the 95% confidence interval did not include 1.0. Relative risks are adjusted for sex, age, and calendar period.

significantly increased even if the last infection occurred more than 15 years previously.

The combined effect of autoimmune diseases and infections on the risk of developing schizophrenia spectrum disorders

The relationship between autoimmune disease and infections as risk factors for schizophrenia has only been investigated in the Danish study. Separating the effect of infections diminished the effect of autoimmune diseases from 45 to 29% on the risk of developing schizophrenia, whereas autoimmune diseases did not influence the 60% increased risk of schizophrenia related to hospital contacts with infections.[39] When a person had both an autoimmune diagnosis and hospital contact with an infection, the risk of schizophrenia increased by 2.25 times (Table 2). The synergy between autoimmune diseases and infections on the risk of schizophrenia was significant, indicating that the effect is larger than would be predicted under an additive model (but note that the multiplicative interaction was not statistically significant). A significant dose–response relationship was found among autoimmune disease, number of severe infections, and risk of schizophrenia (Fig. 1). The risk of schizophrenia increased 3.4 times in persons with an autoimmune disease and hospital contact with three or more infections. The results remained significant after excluding persons diagnosed with substance use disorders and there were no important differences in the relative risk of persons with or without a psychiatric family history. Persons with autoimmune hepatitis and infections had a ninefold increased risk of schizophrenia, and the risk similarly increased if a person had both a hepatitis infection and an autoimmune disease. The association between inflammatory diseases in the liver and schizophrenia[39] is interesting because autoimmune hepatitis is also associated with brain-reactive antibodies[65] and the severe effects on the liver seen in coma hepaticum patients can be dominated by psychiatric symptoms in the initial phases. However, psychiatric symptoms associated with inflammation in the liver could also be explained by the effect of metabolic syndrome or substance abuse. Persons with sepsis infection and an autoimmune disease had a fivefold increased risk of schizophrenia, and sepsis is the type of infection that would

probably increase BBB permeability the most. Interestingly, recent CSF screening studies of patients with schizophrenia, and no known autoimmune diseases or infection, have detected autoantibodies or antibodies against infectious agents in the CSF of 3.2–6% of patients with schizophrenia.[17,66]

Possible etiological mechanisms

The underlying etiological mechanisms of the associations between immunological exposure and schizophrenia might be numerous and speculative, but they are not mutually exclusive. However, it is important to remember that most persons with infections or autoimmune diseases do not develop symptoms of schizophrenia, yet it is possible that exposures to immune challenges and other potential stressors in genetically susceptible individuals may increase the risk of schizophrenia. It is interesting, to note that a similar etiological mechanism is hypothesized for the initiation of autoimmunity, in which genetic susceptibility is required along with triggering events, such as infection or toxin exposure.[38]

Low-grade inflammation and mild encephalitis have been proposed as an underlying causal mechanism of subgroups of patients with schizophrenia.[67] Some infectious agents have the potential to penetrate the BBB and invade the CNS directly, possibly after reaching a threshold level,[68] whereas others cause immune-mediated CNS disorders that may develop through a mechanism involving molecular mimicry.[69] Infections, such as *Streptococcus*, can induce an autoimmune response with molecular mimicry,[14,70] which is considered to induce pediatric autoimmune neuropsychiatric disorders.[12,71] Antibodies to several infectious agents have been demonstrated to cross-react with human neural tissue.[72] Brain-reactive antibodies are better known in connection with paraneoplastic syndromes, in which they have been associated with neurological and psychiatric symptoms.[73] In patients with limbic encephalitis, the most dominant symptoms are often psychiatric symptoms in the initial and remission phases.[74] Animal studies have shown that if brain-reactive antibodies are present in the blood, and agents that increase the permeability of the BBB are given, there is an influx of brain-reactive antibodies into the brain and a subsequent development of a neuropsychiatric syndrome.[75] This indicates that brain-reactive antibodies in

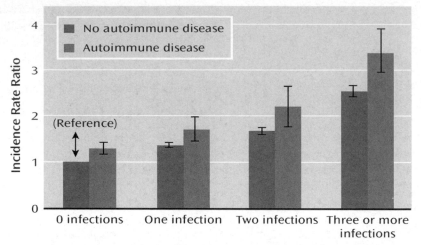

Figure 1. Incidence rate ratios, with confidence intervals, of schizophrenia spectrum disorders in persons with autoimmune disease and infections. The linear trend between number of infections is significant (0, 1, 2, 3+).

the circulation might not have pathological consequences until there is a breach of BBB integrity.[70] However, activated B cells have also been observed to cross the intact BBB and to clonally expand and produce antibodies in the CNS.[76,77] In autoimmune diseases such as systemic lupus erythematosus, between 14 and 75% of individuals experience neuropsychiatric symptoms,[7] which are suspected to be induced by brain-reactive antibodies[13] with affinity to the *N*-methyl-D-aspartate (NMDA) glutamate receptor in the brain,[14,42,78] a receptor that is central to current pathophysiological theories of schizophrenia.[41,79]

Subgroups of patients with schizophrenia may demonstrate features of an autoimmune process, and the hypothesis is strengthened by the findings of an increased prevalence of autoimmune disease and infections in relatives of patients with schizophrenia.[11,47,80] The association of schizophrenia with a range of autoimmune diseases may reflect a general activation of the innate immune system, which somehow affects the brain possibly through increased activity of the hypothalamic-pituitary-adrenal axis.[81] Because of the extremely high comorbidity of autoimmune diseases,[82] association with many diseases could actually reflect a relationship to just one or a few. The candidate causal disease might be celiac disease, for instance, which is associated with many other autoimmune diseases in several studies.[82,83] This possibility implicates the adaptive immune system with breakdown of

self-tolerance and the permeability of the gut in the etiologic pathway to schizophrenia.

The results in Table 3 showing increased risk for schizophrenia for a range of sites of infections suggests that the increase in risk might be due to increased general inflammation. The increased inflammation in both autoimmune diseases and infections may influence the brain through increased BBB permeability, making the brain vulnerable to immune components, such as autoantibodies and cytokines, or possibly to the effect of specific T cell subsets, which are involved in immune surveillance of the brain.[84] Peripheral inflammation can also affect the brain without passing the BBB by proinflammatory cytokines activating the tryptophan–kynurenine pathway that regulates NMDA glutamate receptor activity together with serotonin production[81] and may indirectly affect dopamine regulation.[2] Animal models have indicated that CNS infections during early life can activate the entire kynurenine pathway.[85] Studies have also indicated that peripheral inflammation can affect the brain through stimulation of peripheral nerves such as the vagal nerve.[81] Peripheral inflammation and infection can affect the brain microglia, which arise from monocyte subpopulations and participate in neuronal signal transduction.[86] Inflammation might act as a priming event on microglia, inducing a long-term development of abnormal signal patterns possibly involved in schizophrenia.[87] Activated microglia have been observed in studies of postmortem brains of patients with schizophrenia[88,89] and, recently, *in vivo* utilizing brain imaging of patients with recent-onset schizophrenia.[90] There are also reports of activated lymphocytes in the CSF of patients with acute schizophrenia,[3] and peripheral infections have been shown to affect cognition of persons with ongoing neuroinflammation.[91] Psychological stress has also been proposed as a risk factor for schizophrenia[92] and might contribute to the immunological pathways by increasing the risk of acquiring infections and enhancing immunological responses.[93]

The findings of increased incidence of autoimmune diseases and infections in persons not yet diagnosed with schizophrenia might also be an epiphenomenon related to other risk factors for schizophrenia.[46] Genetic associations between the psychiatric and autoimmune disorders may increase the prevalence in these patients. However, when the analysis was stratified, no additional increase of risk as observed in the persons with a psychiatric family history.[39] Other potential risk factors include infections during maternal pregnancy that can permanently alter the peripheral immune system of the fetus and thereby might change the response and vulnerability to infections.[94] In addition, inflammation and severe infections may reactivate less severe infections that have been associated with schizophrenia, such as herpes virus, toxoplasma, or *Chlamydophila* infections. Some infectious agents escape surveillance by the immune system, but after an acute inflammation or infection, symptoms may flare up from the latent infection.[60] Detection bias seems unlikely to explain the entire association because the increased risk of schizophrenia was still significant even if the last infection occurred more than 15 years previously[39] and even if the diagnosis of an autoimmune disease occurred more than 5 years previously.[33] The results could be influenced by a delay in seeking help in persons not yet diagnosed with schizophrenia, but previous studies indicate that patients with schizophrenia are, in fact, suffering from undertreatment of somatic comorbidity, which may explain the increased mortality of those with schizophrenia.[95] Socioeconomic factors and an increased probability of substance abuse in persons not yet diagnosed with schizophrenia might additionally affect the tendency to acquire severe infections. Furthermore, alcohol and drug abuse can suppress the immune system and thereby increase the vulnerability to acquire infections.[12] Medical treatment against some autoimmune diseases could increase the risk of psychosis; however, a decreased risk of psychosis associated with use of steroids has also been reported.[96] There is no evidence of antibiotics being related to an increased risk of psychosis. It is more likely that the effect is to reduce risk.[97]

Conclusion and perspectives

Autoimmune diseases and severe infections are independent as well as synergistic risk factors for schizophrenia. Hospital contacts for infections occurred in nearly 24% of all patients before a diagnosis of schizophrenia and may be an important risk factor, amounting to a population-attributable risk of 9%. The site of the infection did not have much influence on the increased risk, and the findings are compatible with the hypothesis of a general immunological response affecting the brain. Studies have suggested that antipsychotic drugs may suppress the replication of some infectious agents,[54,98] and interesting results with anti-inflammatory add-on treatment or immunotherapy in patients with schizophrenia have been demonstrated.[99]

Environmental influences such as infections may interact with genetic factors, and a possible association between the age periods during which infections occur needs to be investigated in prospective studies together with the relationship to different genetic markers.[100] New brain-reactive antibodies have been identified in recent years,[42] and other yet unidentified brain-reactive antibodies or pathogens could prove to have a contributing role in the etiology of severe mental illnesses. Although the hypothesis of an immunological contribution to the development of schizophrenia is interesting, it remains unclear precisely how the immunological process affects the brain; whether it is a causal relationship or an epiphenomenon due to genetics or psychological stress needs further investigation. However, the evidence now suggests that autoimmune processes may be involved in the prodrome and perhaps etiology of a considerable proportion of persons with schizophrenia. Screening for autoimmune diseases and infections, preferentially with material closer to the brain, such as CSF in addition to sera, may prove helpful in the diagnosis and treatment planning for persons with first-onset symptoms of schizophrenia, possibly before initiation of antipsychotic treatment. A clear recommendation for screening and specific treatment would be justified only after positive results from well-conducted, randomized clinical trials.

Acknowledgments

This work was supported by a grant from the Stanley Medical Research Institute. W. Eaton's work was supported by NIMH Grant R01 MH53188.

Conflicts of interest

The authors declare no conflicts of interest.

References

1. Mortensen, P.B., M.G. Pedersen & C.B. Pedersen. 2010. Psychiatric family history and schizophrenia risk in Denmark: which mental disorders are relevant? *Psychol. Med.* **40:** 201–210.
2. Muller, N. & M.J. Schwarz. 2010. Immune system and schizophrenia. *Curr. Immunol. Rev.* **6:** 213–220.
3. Nikkila, H.V., K. Muller, A. Ahokas, *et al.* 2001. Increased frequency of activated lymphocytes in the cerebrospinal fluid of patients with acute schizophrenia. *Schizophr. Res.* **49:** 99–105.
4. Potvin, S., E. Stip, A.A. Sepehry, *et al.* 2008. Inflammatory cytokine alterations in schizophrenia: a systematic quantitative review. *Biol. Psychiatry* **63:** 801–808.
5. Marballi, K., M.P. Quinones, F. Jimenez, *et al.* 2010. In vivo and in vitro genetic evidence of involvement of neuregulin 1 in immune system dysregulation. *J. Mol. Med. (Berl)* **88:** 1133–1141.
6. Stefansson, H., R.A. Ophoff, S. Steinberg, *et al.* 2009. Common variants conferring risk of schizophrenia. *Nature* **460:** 744–747.
7. Jones, A.L., B.J. Mowry, M.P. Pender & J.M. Greer. 2005. Immune dysregulation and self-reactivity in schizophrenia: do some cases of schizophrenia have an autoimmune basis? *Immunol. Cell Biol.* **83:** 9–17.
8. Dalman, C., P. Allebeck, D. Gunnell, *et al.* 2008. Infections in the CNS during childhood and the risk of subsequent psychotic illness: a cohort study of more than one million Swedish subjects. *Am. J. Psychiatry* **165:** 59–65.
9. Yolken, R.H. & E.F. Torrey. 2008. Are some cases of psychosis caused by microbial agents? A review of the evidence. *Mol. Psychiatry* **13:** 470–479.
10. Niebuhr, D.W., A.M. Millikan, D.N. Cowan, *et al.* 2008. Selected infectious agents and risk of schizophrenia among U.S. military personnel. *Am. J. Psychiatry* **165:** 99–106.
11. Eaton, W.W., M. Byrne, H. Ewald, *et al.* 2006. Association of schizophrenia and autoimmune diseases: linkage of Danish national registers. *Am. J. Psychiatry* **163:** 521–528.
12. Margutti, P., F. Delunardo & E. Ortona. 2006. Autoantibodies associated with psychiatric disorders. *Curr. Neurovasc. Res.* **3:** 149–157.
13. Ballok, D.A. 2007. Neuroimmunopathology in a murine model of neuropsychiatric lupus. *Brain Res. Rev.* **54:** 67–79.
14. Rice, J.S., C. Kowal, B.T. Volpe, *et al.* 2005. Molecular mimicry: anti-DNA antibodies bind microbial and nonnucleic acid self-antigens. *Curr. Top. Microbiol. Immunol.* **296:** 137–151.
15. Irani, S. & B. Lang. 2008. Autoantibody-mediated disorders of the central nervous system. *Autoimmunity* **41:** 55–65.
16. Uranova, N.A., I.S. Zimina, O.V. Vikhreva, *et al.* 2010. Ultrastructural damage of capillaries in the neocortex in schizophrenia. *World J. Biol. Psychiatry* **11:** 567–578.
17. Bechter, K., H. Reiber, S. Herzog, *et al.* 2010. Cerebrospinal fluid analysis in affective and schizophrenic spectrum disorders: identification of subgroups with immune responses and blood-CSF barrier dysfunction. *J. Psychiatr. Res.* **44:** 321–330.
18. Laske, C., M. Zank, R. Klein, *et al.* 2008. Autoantibody reactivity in serum of patients with major depression, schizophrenia and healthy controls. *Psychiatry Res.* **158:** 83–86.
19. Tanaka, S., H. Matsunaga, M. Kimura, *et al.* 2003. Autoantibodies against four kinds of neurotransmitter receptors in psychiatric disorders. *J. Neuroimmunol.* **141:** 155–164.
20. Trevathan, R. & J.C. Tatum. 1953. Rarity of concurrence of psychosis and rheumatoid arthritis in individual patients. *J. Nerv. Ment. Dis.* **120:** 83–84.
21. Pilkington, T. 1955. The coincidence of rheumatoid arthritis and schizophrenia. *J. Nerv. Ment. Dis.* **124:** 604–606.
22. Eaton, W.W. & C.-Y. Chen. 2006. Epidemiology. In *The American Psychiatric Publishing Textbook of Schizophrenia.* J.A. Lieberman, T.S. Stroup & D.O. Perkins, Eds.: 17–38. American Psychiatric Publishing. Washington, D.C.
23. Bender, L. 1953. Childhood schizophrenia. *Psychiatr. Q.* **27:** 663–681.
24. Graff, H. & A. Handford. 1961. Celiac Syndrome in the case histories of five schizophrenics. *Psychiatr. Q.* **35:** 306–313.
25. Kalaydjian, A.E., W. Eaton, N. Cascella & A. Fasano. 2006. The gluten connection: the association between schizophrenia and celiac disease. *Acta. Psychiatr. Scand.* **113:** 82–90.
26. Reichelt, K.L. & J. Landmark. 1995. Specific IgA antibody increases in schizophrenia. *Biol. Psychiatry* **37:** 410–413.
27. Cascella, N.G., D. Kryszak, B. Bhatti, *et al.* 2011. Prevalence of celiac disease and gluten sensitivity in the United States clinical antipsychotic trials of intervention effectiveness study population. *Schizophr. Bull.* **37:** 94–100.
28. Jin, S.Z., N. Wu, Q. Xu, *et al.* 2012. A study of circulating gliadin antibodies in schizophrenia among a Chinese population. *Schizophr. Bull.* **38:** 514–518.
29. Samaroo, D., F. Dickerson, D.D. Kasarda, *et al.* 2010. Novel immune response to gluten in individuals with schizophrenia. *Schizophr. Res.* **118:** 248–255.
30. Dickerson, F., C. Stallings, A. Origoni, *et al.* 2010. Markers of gluten sensitivity and celiac disease in recent-onset psychosis and multi-episode schizophrenia. *Biol. Psychiatry* **68:** 100–104.
31. Hadjivassiliou, M., D.S.G.R.A. Sanders, N. Woodroofe, *et al.* 2010. Gluten sensitivity: from gut to brain. *Lancet Neurology* **9:** 318–330.

32. Jackson, J.R., W.W. Eaton, N.G. Cascella, *et al.* 2012. Neurologic and psychiatric manifestations of celiac disease and gluten sensitivity. *Psychiatr. Q* **83:** 91–102.

33. Eaton, W.W., M.G. Pedersen, P.R. Nielsen & P.B. Mortensen. 2010. Autoimmune diseases, bipolar disorder, and non-affective psychosis. *Bipolar Disord.* **12:** 638–646.

34. Tsai, H., S. Chen, Y. Chao, *et al.* 2012. Prevalence of autoimmune diseases in hospitalized schizophrenic patients: a nationwide population-based study. *Br. J. Psychiatry* **200:** 374–380.

35. Fessel, W.J. 1962. Autoimmunity and mental illness. A preliminary report. *Arch. Gen. Psychiatry* **6:** 320–323.

36. Heath, R.G. & I.M. Krupp. 1967. Schizophrenia as an immunologic disorder. I. Demonstration of antibrain globulins by fluorescent antibody techniques. *Arch. Gen. Psychiatry* **16:** 1–9.

37. Heath, R.G. & I.M. Krupp. 1967. Catatonia induced in monkeys by antibrain antibody. *Am. J. Psychiatry* **123:** 1499–1504.

38. Goldsmith, C.A. & D.P. Rogers. 2008. The case for autoimmunity in the etiology of schizophrenia. *Pharmacotherapy* **28:** 730–741.

39. Benros, M.E., P.R. Nielsen, M. Nordentoft, *et al.* 2011. Autoimmune diseases and severe infections as risk factors for schizophrenia. A 30-year population-based register study. *Am. J. Psychiatry.* **168:** 1303–1310.

40. Schott, K., J.E. Schaefer, E. Richartz, *et al.* 2003. Autoantibodies to serotonin in serum of patients with psychiatric disorders. *Psychiatry Res.* **121:** 51–57.

41. Zandi, M.S., S.R. Irani, B. Lang, *et al.* 2011. Disease-relevant autoantibodies in first episode schizophrenia. *J. Neurol.* **258:** 686–688.

42. Graus, F., A. Saiz & J. Dalmau. 2010. Antibodies and neuronal autoimmune disorders of the CNS. *J. Neurol.* **257:** 509–517.

43. Noll, R. 2007. Kraepelin's 'lost biological psychiatry'? Autointoxication, organotherapy and surgery for dementia praecox. *Hist. Psychiatry* **18:** 301–320.

44. Torrey, E.F., M.F. Leweke, M.J. Schwarz, *et al.* 2006. Cytomegalovirus and schizophrenia CNS. *Drugs* **20:** 879–885.

45. Pedersen, C.B. & P.B. Mortensen. 2001. Evidence of a dose-response relationship between urbanicity during upbringing and schizophrenia risk. *Arch. Gen. Psychiatry* **58:** 1039–1046.

46. Brown, A.S. & E.J. Derkits. 2010. Prenatal infection and schizophrenia: a review of epidemiologic and translational studies. *Am. J. Psychiatry* **167:** 261–280.

47. Nielsen, P.R., T.M. Laursen & P.B. Mortensen. 2011. Association between parental hospital-treated infection and the risk of schizophrenia in adolescence and early adulthood. *Schizophr. Bull.* Oct 20. [Epub ahead of print].

48. Arias, I., A. Sorlozano, E. Villegas, *et al.* 2012. Infectious agents associated with schizophrenia: a meta-analysis. *Schizophr. Res.* **136:** 128–136.

49. Torrey, E.F., J.J. Bartko, Z.R. Lun & R.H. Yolken. 2007. Antibodies to *Toxoplasma gondii* in patients with schizophrenia: a meta-analysis. *Schizophr. Bull.* **33:** 729–736.

50. Pedersen, M.G., H. Stevens, C.B. Pedersen, *et al.* 2011. Toxoplasma infection and later development of schizophrenia in mothers. *Am. J. Psychiatry* **168:** 814–821.

51. Dickerson, F.B., J.J. Boronow, C. Stallings, *et al.* 2003. Association of serum antibodies to herpes simplex virus 1 with cognitive deficits in individuals with schizophrenia. *Arch. Gen. Psychiatry* **60:** 466–472.

52. Niebuhr, D.W., A.M. Millikan, R. Yolken, *et al.* 2008. Results from a hypothesis generating case-control study: herpes family viruses and schizophrenia among military personnel. *Schizophr. Bull.* **34:** 1182–1188.

53. Bartova, L., J. Rajcani & J. Pogady. 1987. Herpes simplex virus antibodies in the cerebrospinal fluid of schizophrenic patients. *Acta Virol.* **31:** 443–446.

54. Leweke, F.M, C.W. Gerth, D. Koethe, *et al.* 2004. Antibodies to infectious agents in individuals with recent onset schizophrenia. *Eur. Arch. Psychiatry Clin. Neurosci.* **254:** 4–8.

55. Torrey, E.F., R.H. Yolken & C.J. Winfrey. 1982. Cytomegalovirus antibody in cerebrospinal fluid of schizophrenic patients detected by enzyme immunoassay. *Science* **216:** 892–894.

56. Karlsson, H., S. Bachmann, J. Schroder, *et al.* 2001. Retroviral RNA identified in the cerebrospinal fluids and brains of individuals with schizophrenia. *Proc. Natl. Acad. Sci. U.S.A.* **98:** 4634–4639.

57. Hart, D.J., R.G. Heath, F.J. Sautter, Jr., *et al.* 1999. Antiretroviral antibodies: implications for schizophrenia, schizophrenia spectrum disorders, and bipolar disorder. *Biol. Psychiatry* **45:** 704–714.

58. Chen, C.H., Y.L. Chiu, F.C. Wei, *et al.* 1999. High seroprevalence of Borna virus infection in schizophrenic patients, family members and mental health workers in Taiwan. *Mol. Psychiatry* **4:** 33–38.

59. Fellerhoff, B., B. Laumbacher, N. Mueller, *et al.* 2007. Associations between Chlamydophila infections, schizophrenia and risk of HLA-A10. *Mol. Psychiatry* **12:** 264–272.

60. Fellerhoff, B. & R. Wank. 2011. Increased prevalence of Chlamydophila DNA in post-mortem brain frontal cortex from patients with schizophrenia. *Schizophr. Res.* **129:** 191–195.

61. Koponen, H., P. Rantakallio, J. Veijola, *et al.* 2004. Childhood central nervous system infections and risk for schizophrenia. *Eur. Arch. Psychiatry Clin. Neurosci.* **254:** 9–13.

62. Abrahao, A.L., R. Focaccia & W.F. Gattaz. 2005. Childhood meningitis increases the risk for adult schizophrenia. *World J. Biol. Psychiatry* **6**(Suppl. 2): 44–48.

63. Weiser, M., N. Werbeloff, A. Levine, *et al.* 2010. CNS infection in childhood does not confer risk for later schizophrenia: a case-control study. *Schizophr. Res.* **124:** 231–235.

64. Krause, D., J. Matz, E. Weidinger, *et al.* 2010. The association of infectious agents and schizophrenia. *World J. Biol. Psychiatry* **11:** 739–743.

65. Kimura, A., T. Sakurai, A. Koumura, *et al.* 2010. High prevalence of autoantibodies against phosphoglycerate mutase 1 in patients with autoimmune central nervous system diseases. *J. Neuroimmunol.* **219:** 105–108.

66. Kranaster, L., D. Koethe, C. Hoyer, *et al.* 2011. Cerebrospinal fluid diagnostics in first-episode schizophrenia. *Eur. Arch. Psychiatry Clin. Neurosci.* **261:** 529–530.

67. Bechter, K. 2004. [The mild encephalitis-hypothesis–new findings and studies]. *Psychiatr. Prax.* **31**(Suppl. 1): S41–S43.

68. Kim, K.S. 2008. Mechanisms of microbial traversal of the blood-brain barrier. *Nat. Rev. Microbiol.* **6:** 625–634.

69. Rose, N.R. 1998. The role of infection in the pathogenesis of autoimmune disease. *Semin. Immunol.* **10:** 5–13.

70. Diamond, B., P.T. Huerta, P. Mina-Osorio, *et al.* 2009. Losing your nerves? Maybe it's the antibodies. *Nat. Rev. Immunol.* **9:** 449–456.

71. Swedo, S.E., H.L. Leonard, M. Garvey, *et al.* 1998. Pediatric autoimmune neuropsychiatric disorders associated with streptococcal infections: clinical description of the first 50 cases. *Am. J. Psychiatry* **155:** 264–271.

72. Birner, P., B. Gatterbauer, D. Drobna & H. Bernheimer. 2000. Molecular mimicry in infectious encephalitis and neuritis: binding of antibodies against infectious agents on Western blots of human nervous tissue. *J. Infect.* **41:** 32–38.

73. Kayser, M.S., C.G. Kohler & J. Dalmau. 2010. Psychiatric manifestations of paraneoplastic disorders. *Am. J. Psychiatry* **167:** 1039–1050.

74. Kayser, M.S. & J. Dalmau. 2011. Anti-NMDA receptor encephalitis in psychiatry. *Curre. Psychiatry Rev.* **7:** 189–193.

75. Kowal, C., L.A. Degiorgio, T. Nakaoka, *et al.* 2004. Cognition and immunity: antibody impairs memory. *Immunity* **21:** 179–188.

76. Dalakas, M.C. 2008. Invited article: inhibition of B cell functions: implications for neurology. *Neurology* **70:** 2252–2260.

77. Von Budingen, H.C., A. Bar-Or & S.S. Zamvil. 2011. B cells in multiple sclerosis: connecting the dots. *Curr. Opin. Immunol.* **23:** 713–720.

78. Degiorgio, L.A., K.N. Konstantinov, S.C. Lee, *et al.* 2001. A subset of lupus anti-DNA antibodies cross-reacts with the NR2 glutamate receptor in systemic lupus erythematosus. *Nat. Med.* **7:** 1189–1193.

79. Balu, D.T. & J.T. Coyle. 2011. Neuroplasticity signaling pathways linked to the pathophysiology of schizophrenia. *Neurosci. Biobehav. Rev.* **35:** 848–870.

80. Wright, P., P.C. Sham, C.M. Gilvarry, *et al.* 1996. Autoimmune diseases in the pedigrees of schizophrenic and control subjects. *Schizophr. Res.* **20:** 261–267.

81. Dantzer, R., J.C. O'Connor, G.G. Freund, *et al.* 2008. From inflammation to sickness and depression: when the immune system subjugates the brain. *Nat. Rev. Neurosci.* **9:** 46–56.

82. Eaton, W.W., N.R. Rose, A. Kalaydjian, *et al.* 2007. Epidemiology of autoimmune diseases in Denmark. *J. Autoimmun.* **29:** 1–9.

83. James, M. & B. Scott. 2001. Coeliac disease: the cause of the various associated disorders? *Eur. J. Gastroenterol. Hepatol.* **13:** 1119–1121.

84. Goverman, J. 2009. Autoimmune T cell responses in the central nervous system. *Nat. Rev. Immunol.* **9:** 393–407.

85. Holtze, M., L. Asp, L. Schwieler, *et al.* 2008. Induction of the kynurenine pathway by neurotropic influenza A virus infection. *J. Neurosci. Res.* **86:** 3674–3683.

86. Kaur, C., S.T. Dheen & E.A. Ling. 2007. From blood to brain: amoeboid microglial cell, a nascent macrophage and its functions in developing brain. *Acta Pharmacol. Sin.* **28:** 1087–1096.

87. Hickie, I.B., R. Banati, C.H. Stewart & A.R. Lloyd. 2009. Are common childhood or adolescent infections risk factors for schizophrenia and other psychotic disorders? *Med. J. Aust.* **190:** S17–S21.

88. Bayer, T.A., R. Buslei, L. Havas & P. Falkai. 1999. Evidence for activation of microglia in patients with psychiatric illnesses. *Neurosci. Lett.* **271:** 126–128.

89. Radewicz, K., L.J. Garey, S.M. Gentleman & R. Reynolds. 2000. Increase in HLA-DR immunoreactive microglia in frontal and temporal cortex of chronic schizophrenics. *J. Neuropathol. Exp. Neurol.* **59:** 137–150.

90. Van Berckel, B.N., M.G. Bossong, R. Boellaard, *et al.* 2008. Microglia activation in recent-onset schizophrenia: a quantitative (R)-[11C]PK11195 positron emission tomography study. *Biol. Psychiatry* **64:** 820–822.

91. Perry, V.H., C. Cunningham & C. Holmes. 2007. Systemic infections and inflammation affect chronic neurodegeneration. *Nat. Rev. Immunol.* **7:** 161–167.

92. van Winkel R., N.C. Stefanis & I. Myin-Germeys. 2008. Psychosocial stress and psychosis. A review of the neurobiological mechanisms and the evidence for gene-stress interaction. *Schizophr. Bull.* **34:** 1095–1105.

93. Pedersen, A., R. Zachariae & D.H. Bovbjerg. 2010. Influence of psychological stress on upper respiratory infection—a meta-analysis of prospective studies. *Psychosom. Med.* **72:** 823–832.

94. Patterson, P.H. 2009. Immune involvement in schizophrenia and autism: etiology, pathology and animal models. *Behav. Brain Res.* **204:** 313–321.

95. Laursen, T.M., T. Munk-Olsen & C. Gasse. 2011. Chronic somatic comorbidity and excess mortality due to natural causes in persons with schizophrenia or bipolar affective disorder. *PLoS One* **6:** e24597.

96. Laan, W., H. Smeets, N.J. De Wit, *et al.* 2009. Glucocorticosteroids associated with a decreased risk of psychosis. *J. Clin. Psychopharmacol.* **29:** 288–290.

97. Frykholm, B.O. 2009. On the question of infectious aetiologies for multiple sclerosis, schizophrenia and the chronic fatigue syndrome and their treatment with antibiotics. *Med. Hypotheses* **72:** 736–739.

98. Jones-Brando, L., E.F. Torrey & R. Yolken. 2003. Drugs used in the treatment of schizophrenia and bipolar disorder inhibit the replication of *Toxoplasma gondii*. *Schizophr. Res.* **62:** 237–244.

99. Meyer, U., M.J. Schwarz & N. Muller. 2011. Inflammatory processes in schizophrenia: a promising neuroimmunological target for the treatment of negative/cognitive symptoms and beyond. *Pharmacol. Ther.* **132:** 96–110.

100. Carter, C.J. 2009. Schizophrenia susceptibility genes directly implicated in the life cycles of pathogens: cytomegalovirus, influenza, herpes simplex, rubella, and *Toxoplasma gondii*. *Schizophr. Bull.* **35:** 1163–1182.

Ann. N.Y. Acad. Sci. ISSN 0077-8923

ANNALS OF THE NEW YORK ACADEMY OF SCIENCES
Issue: *Neuroimmunomodulation in Health and Disease*

Antidepressants prevent hierarchy destabilization induced by lipopolysaccharide administration in mice: a neurobiological approach to depression

Daniel Wagner Hamada Cohn, Denise Kinoshita, and João Palermo-Neto

Neuroimmunomodulation Research Group, Department of Pathology, School of Veterinary Medicine, University of São Paulo, Brazil

Address for correspondence: Daniel Wagner Hamada Cohn, Dr. Orlando Marques de Paiva, 87 ZIP Code 05508-270, Cidade Universitária, São Paulo/SP Brazil. dcohn@usp.br; dwhcohn@gmail.com

In spite of the high prevalence and negative impact of depression, little is known about its pathophysiology. Basic research on depression needs new animal models in order to increase knowledge of the disease and search for new therapies. The work presented here aims to provide a neurobiologically validated model for investigating the relationships among sickness behavior, antidepressants treatment, and social dominance behavior. For this purpose, dominant individuals from dyads of male Swiss mice were treated with the bacterial endotoxin lipopolysaccharide (LPS) to induce social hierarchy destabilization. Two groups were treated with the antidepressants imipramine and fluoxetine prior to LPS administration. In these groups, antidepressant treatment prevented the occurrence of social destabilization. These results indicate that this model could be useful in providing new insights into the understanding of the brain systems involved in depression.

Keywords: depression models; lipopolysaccharide; social behavior; social hierarchies; antidepressants

Introduction

Depression is a severe psychiatric syndrome characterized by a set of affective, cognitive, and neurovegetative symptoms that often run a relapse and remitting course over long periods of time.[1] Its core symptoms are depressed mood, anhedonia, irritability (affective symptoms), difficulties in concentrating (cognitive symptoms), and abnormalities in appetite and sleep (neurovegetative symptoms).[2]

This disease is a major concern for personal and economic welfare and it is the fourth leading cause of disability worldwide. According to the World Health Organization (WHO), depression has greater negative impact on quality of life when compared with cardiovascular disease and is projected to be the leading cause of disability by the year 2030.[3]

Despite the high prevalence and considerable social impact of depression, little is known about its pathophysiology, especially compared with other common relevant multifactorial conditions.[4] Moreover, no treatments are able to target the causal factors of the disease.[5] Less then 60% of depressive patients achieve remission with current drug therapies.[6]

This discrepancy could be explained by two factors. Regarding clinical research, it is not easy to analyze pathological changes within the brain of depressed patients. As for basic research, there are some difficulties in modeling depression using laboratory animals.[4,7] The lack of emphasis on translational research is another issue pertaining to these two factors.[8] Translational research is conceptualized here according to the bench-to-bedside definition: the dialog between epidemiological/clinical investigation and basic research.[9]

The aim of the present paper is to provide an alternative approach to the basic research aspect of depression: one that is primarily concerned with neurobiological validation. Moreover, an ethological and neuroimmunomodulatory perspective will also be considered. For this purpose, a brief overview

doi: 10.1111/j.1749-6632.2012.06635.x

on the hypothesis of depression mechanisms will be provided. Emphasis will be given to the inflammatory hypothesis. A nonanthropomorphic perspective on depression models will then be discussed. This perspective, which would benefit from ethological methods, will be exemplified by a few experimental results from our laboratory.

Mechanisms and the cytokine hypothesis

The implication of serotonin (5-HT) in the pathophysiology of depression dates back over 40 years, when it was discovered that the first generation of antidepressants blocked 5-HT (serotonin) reuptake and metabolism as part of their pharmacological effects. The link between 5-HT and depression has continued to provide important discoveries. For example, the main class of drugs used in therapy today are the selective serotonin reuptake inhibitors (SSRIs).[1] Another important finding that came from clinical research was the observation that cortisol secretion was elevated in depressive patients.[10] It is now believed that changes in 5-HT and cortisol are a result of stress in depression rather than a causal factor in the etiology of the disease.[11]

Response of the serotonergic system to antidepressant administration seems to be one component of the mechanisms involved in clinically relevant neurological changes. As a general rule, antidepressants increase the availability of 5-HT in the synaptic cleft. Currently, it is believed that increased synaptic 5-HT leads to enhanced neurogenesis, which is impaired in depression.[12] This increase also reduces neurodegeneration, which seems to also play an important role.[1] In this context, neurotrophins, such as brain-derived neurotrophic factor (BDNF), have been shown to be reduced in stress-induced animal models of depression.[13]

Other evidence indicates that inflammatory processes are involved in the pathogenesis of depression.[14–16] This hypothesis is termed the *cytokine hypothesis* of depression. Clinical evidence shows that patients with major depression display increased levels of proinflammatory cytokines such as interleukin-1 (IL-1), IL-6, interferon-γ (INF-γ), and tumor necrosis factor-α (TNF-α).[17] Other inflammatory markers related to the acute phase response, such as the acute phase proteins, and hypozincemia have also been linked to depression.[18,19]

These pieces of evidence, corroborated by research with laboratory animals (for a review, see Ref. 20), bring neuroimmunomodulation (or psychoneuroimmunology) into the scope of depression research.

From a behavioral science perspective, administration of proinflammatory cytokines in rodents promotes behavioral changes (e.g., anorexia, enhanced slow-wave sleep, decreased social contacts, and anhedonia) that are collectively referred to as *sickness behavior*.[20] These behavioral changes follow acute infection or inflammation and are part of the fever syndrome. Additionally, these changes provide optimal conditions to mount the fever and immunological responses needed to overcome infection.[21] It is now well known that pathogen-associated molecular patterns (PAMPs) such as lipopolysaccharide (LPS), a component of the cell wall of Gram-negative bacteria, can induce the secretion of proinflammatory cytokines leading to sickness behavior.[22]

Taken together, this evidence indicates that sickness behavior and depression might display overlapping mechanisms.[15,23] Indeed, antidepressant administration has been shown to counteract cytokine and LPS-induced behavioral and physiological changes. These data have been shown in different contexts, including models used to assess sickness behavior parameters (e.g., fever), [24] behavioral models traditionally used in depression research (e.g., forced swimming), [25] the secretion of proinflammatory cytokines,[26] and in specific brain nuclei activation.[27]

However, it is important to mention that, in contrast, other work suggests that the relationship between sickness behavior and depression may be less clear.[28,29]

Challenges in modeling depression

Animal models are ubiquitous in biomedical sciences, and great effort is devoted to developing useful experimental paradigms. In the case of psychopathology research, animal models are confronted with the challenge of modeling symptoms that are subjective feelings and thoughts of humans.[7]

In some cases, this task may be simpler because of the evolutionary correlation (between humans and laboratory rodents) of the behavioral system studied. This is the case in some psychopathologies such as anxiety disorders. When assessing anxiety in rodents, one can assume to be dealing with a behavioral system that bears an evolutionary correlation with man. Anxiety in humans may be a subjective (and potentially maladaptive) feeling, but it is also the expression of a behavioral defense system that is, at least to some degree, shared among different mammals. The same reasoning could be applied to stress research, for example.

However, in the case of depression, such reasoning does not apply, since depression cannot be considered as an expression (or an overexpression) of some evolved behavioral system.[a]

Therefore, modeling depression requires special attention to validation. The limited efficacy of pharmacological therapy in depression indicates the limited scope of predictive or pharmacological validity. In addition, anthropomorphic assumptions also limit the usefulness of mouse models. As a result, face validity also has a limited scope. Regarding construct validity, it is not possible to include in the models all of the aspects (neurobiological, psychosocial, and behavioral) that are required to establish this kind of validity.[7,31]

In order to avoid these issues, it has been proposed that depression models in rodents should have the following two characteristics: they should not make anthropomorphic assumptions, and they should focus on discrete symptoms of the disease (and not try to mimic the whole set of nosological entities of human depression).[7]

Based on these considerations, the work described here proposes a paradigm to study behavioral and brain systems involved in depression. It has been shown that treating the dominant mouse within a dyad with the bacterial endotoxin LPS can disrupt dominance hierarchy.[32,33] In this work, it is shown that if the dominant mouse is treated with LPS, as well as with the antidepressants fluoxetine and imipramine, the hierarchy destabilization can

---

[a]According to the cytokine hypothesis, depression and sickness behavior share some mechanisms. As a result, one could infer that inducing a sickness behavior could model depression. This inference has already been tested and has been ruled out since sickness behavior symptoms resemble those of depression, but are not the same.[30] In addition, sickness behavior in animals already has its counterpart in humans, which is human sickness behavior and not depression.

be prevented. This paradigm does not make anthropomorphic inferences. It uses ethologically validated assumptions, since the behavior under consideration (i.e., intraspecific aggression) is biologically relevant to the experimental subjects. The ultimate goal of this line of work is to provide a neurobiological study of the relation between acute inflammatory states and brain and behavioral systems involved in depression.

The LPS-induced social destabilization paradigm

Materials and methods

Subjects. One hundred and twelve adult Swiss male mice were used. Animals were kept in standard polypropylene cages ($14 \times 30 \times 13$ cm) with wood shaving bedding and a stainless steel grid lid. Ambient temperature was 23 °C with a 12-h light/dark cycle, with lights on at 7:00 a.m. Mice received standard rodent chow and water *ad libitum*.

Animals were housed and used in accordance with the guidelines set forth by the Committee on Care and Use of Laboratory Animals' Resources of the School of Veterinary Medicine at the University of São Paulo.

Reagents. The reagents used in the present experiment included the lipopolysaccharide phenol extract *Escherichia coli* LPS (serotype 0127B6; SIGMA, Saint Louis, Missouri); saline (sterile sodium chloride solution); impipramine hydrochloride (SIGMA); and fluoxetine hydrochloride (SIGMA).

Animal handling and experimental groups

Dyads of mice were assembled immediately after weaning (postnatal day 21) from pups originated from distinct litters. Mice within the pair were chosen according to similarity of size and weight (difference not greater than 1 g). Mice were 80–90 days old when the experiment started. Dyads were assigned to five experimental groups:

Control group. Dominant mice (SAL + LPS; $n = 16$ dyads) were treated with saline followed by LPS administration 1 hour later. Treatment was applied for 3 consecutive days.

Experimental imipramine group. (IMIP + LPS; $n = 10$) dominant mice were treated with imipramine followed by LPS administration 1 hour later. Treatment was applied for 3 consecutive days.

Experimental fluoxetine group. (FLU + LPS; $n = 10$) dominant mice were treated with fluoxetine followed by LPS administration 1 hour later. Treatment was applied for 3 consecutive days.

Control imipramine group. (IMIP + SAL; $n = 10$) dominant mice were treated with imipramine followed by saline administration 1 hour later. Treatment was applied for 3 consecutive days.

Control fluoxetine group. (FLU + SAL; $n = 10$) dominant mice were treated with fluoxetine followed by saline administration 1 hour later. Treatment was applied for 3 consecutive days.

Doses used for LPS were 400 µg/kg. Doses and regimen were chosen based on previous works from our laboratory.[33,34] Doses used for imipramine and fluoxetine were 20 mg/kg and were chosen according to previous work in the literature.[35,36] LPS and the antidepressants were diluted in 0.4 mL of sterile saline and applied intraperitoneally. All the treatments were applied between 11:00 and 13:00 hours.

Social hierarchies assessment

Dominance relationships in the home cages were directly assessed by a trained observer immediately after bedding change, once a week. Cage bedding change removes olfactory cues related to social recognition, thus promoting agonistic interaction in mice.[37,38] Each observation session lasted 40 minutes. If agonistic interaction in the cage was unequivocal (only one mouse displayed offensive behavior while the other displayed only defensive behavior), the experimenter scored *unequivocal dominance* (UD) while identifying which mouse was the dominant. If both mice of the dyad displayed offensive and defensive behavior during the observed agonistic interactions, the experimenter scored *undefined dominance* (ID). *Unequivocal dominance* (UD) could be of two types: *original unequivocal dominance* (OUD) or *inverted unequivocal dominance* (IUD), depending on whether the dominance relationship had changed when comparing pretreatment weeks and posttreatment weeks.

Social hierarchies assessment in the home cages occurred once a week, between 16:00 and 18:00 hours, over 7 weeks (weeks −3, −2, −1, 0, 1, 2, 3). Treatments occurred during the experimental week (week 0). Assessments prior to the experimental week were done to ensure that the dyads displayed stable social hierarchies.

Statistical analysis. Nominal variables in 2×2 contingency tables (treatment × social hierarchies) were analyzed by Fisher's exact test of independence. Nominal variables with dependence (treatment × social hierarchies measured in the same animals measured in different weeks) were compared with McNemar's test.

Results

In all groups, assessment of dominance hierarchies was clear and consistent during the three observations prior to the experimental week (week 0). After bedding change, only one mouse within each dyad displayed offensive behavior, while the other displayed defensive behavior. For statistical analysis purposes destabilization comprised the sum of dyads that scored IUD and ID.

In the control group (SAL + LPS) (Fig. 1), Mc-Nemar's test showed significant differences from assessments in week 0 ($\chi^2 = 7.111$; $P = 0.008$) and week 1 ($\chi^2 = 4.167$; $P = 0.041$) when compared with week −1 (pretreatment week).

In the experimental impipramine group (IMIP + LPS) (Fig. 2A) and the experimental fluoxetine group (FLU + LPS) (Fig. 2B), there were no statistically significant differences between experimental and posttreatment weeks and pretreatment week −1. For instance, Fisher's exact test yielded $P = 1.000$ since only one dyad in each of these

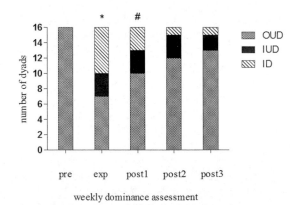

Figure 1. Dominance hierarchies in the control group (SAL + LPS) along assessments in weeks −1, 0 (exp), 1, 2, and 3. OUD, original unequivocal dominance; IUD, inverted unequivocal dominance; ID, undefined dominance. *Statistically significant difference with $P < 0.005$ when compared with pretreatment week; #statistically significant difference with $P < 0.05$ when compared with pretreatment week.

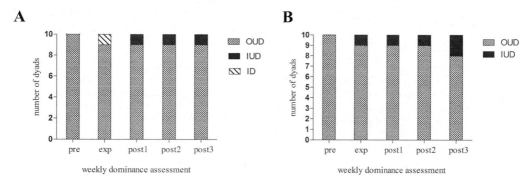

Figure 2. Dominance hierarchies in the impipramine experimental group (IMIP + LPS). (A) The Fluoxetine experimental group (FLU + LPS) (B) along assessments in weeks −1, 0 (exp), 1, 2, and 3. OUD, original unequivocal dominance; IUD, inverted unequivocal dominance; ID, undefined dominance. There are no significant statistical differences among different weekly assessments.

experimental groups displayed hierarchical instability after treatment.

If we further compare week 1 of the experimental groups and week 1 of the control group (SAL + LPS), Fisher's exact test yields statistically significant results ($P = 0.037$). Hierarchies in the impipramine control group (IMIP + SAL) (Fig. 3) and fluoxetine control group (FLU + SAL) remained unchanged during and after the treatment week (Fig. 1).

Discussion

These results show that LPS treatment in the dominant mouse within a dyad destabilizes social hierarchy. Treatment with the antidepressants imipramine and fluoxetine prior to the LPS administration prevented social destabilization. Results from the imipramine and fluoxetine control groups demonstrate that destabilization is an effect of the LPS administration and is not caused by the antidepressant treatment.

In laboratory settings, excessive handling could promote hierarchy destabilization in mice due to increased aggressive arousal.[37] However, previous work[33] with this model has shown that animal handling and intraperitoneal saline injection do not induce social destabilization per se. As a precaution, great care was taken not to introduce stressful environmental variables and all groups and animals were handled in the same manner. Even though social behavior assessment occurred during the light period of the light–dark cycle, when mice are expected to be resting, the animals promptly engaged in social behavior because they were aroused by the bedding change.

Outbred Swiss mice were chosen because of the ethological approach of this work. Genetic heterogeneity provides urine properties heterogeneity.

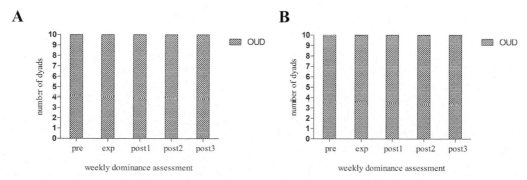

Figure 3. Dominance hierarchies in the imipramine control group (IMIP + SAL). (A) The fluoxetine control group (FLU + SAL) (B) along assessments in weeks −1, 0 (exp), 1, 2, and 3. OUD, original unequivocal dominance. There are no significant statistical differences among different weekly assessments.

Male mice use odor cues secreted in the urine for territory marking.[37] Therefore, odor cue heterogeneity is important if one is working with a more naturalistic approach.

Two classic antidepressants drugs were chosen for this experiment. Fluoxetine is a classic selective serotonin reuptake inhibitor, and imipramine is a classic tryciclic antidepressant that inhibits the reuptake of serotonin as well as noradrenalin.[39] Hierarchies assessment results could not show distinct behavioral effects of each drug. However, it is reasonable to suppose that interesting differences might arise with more detailed behavioral analysis.

LPS-induced sickness behavior is a motivational reorganization,[40] and imipramine and fluoxetine are drugs with antidepressant effects. Therefore, from a neuroethological point of view, these results bring forth one question: How do the antidepressants interact with sickness behavior systems in order to provide motivation for the dominant mouse not to let go of its dominance position? The answer to this question could have implications for depression research, since antidepressants here clearly interact with social aggressive behavior and decision making in an ethologically relevant context.

As stated previously, this work does not aim to offer a new depression paradigm; it aims at providing a model to assess the relation between three factors: (1) behavioral changes induced by LPS administration; (2) behavioral effects of antidepressant drugs; and (3) social and aggressive behavior of male mice. Basically, these factors could be analyzed in the neurobiological level. In that case, the three factors could be (1) brain cytokinergic system activation, (2) brain neurotransmitter systems involved in depression, and (3) brain systems involved in motivation and social agonistic behavior.

Even though the three factors assessed in the paradigm are not directly related to depression, their interaction could provide new insights into some behaviors and brain systems involved in the disease. Preliminary results shown here indicate that this model could be used in depression research. Further studies are required to test its reliability with different compounds, as well as to test its usefulness in assessing the underlying neurological mechanisms.

Acknowledgments

The authors thank FAPESP (Fundação de Amparo a Pesquisa do Estado de São Paulo) for financial support (Processes 09/52419 and 09/51886-3) and Dr. Ilana Gabanyi for helpful comments during the preparation of the manuscript.

Conflicts of interest

The authors declare no conflicts of interest.

References

1. Sharp, T. & P.J. Cowen. 2011. 5-HT and depression: is the glass half-full? *Curr. Opin. Pharmacol.* **11:** 45–51.
2. Nestler, E.J. *et al.* 2002. *Neuron* **34:** 13–25.
3. Mathers, C.D. & D. Loncar. 2006. Projections of global mortality and burden of fDisease from 2002 to 2030. *PLoS ONE* **3:** E442.
4. Krishnan, V. & E.J. Nestler. 2008. The molecular neurobiology of depression. *Nature* **455:** 894–902.
5. Maes, M. *et al.* 2009. The inflammatory & neurodegenerative (I&ND) hypothesis of depression: leads for future research and new drug developments in depression. *Metab. Brain Dis.* **24:** 27–53.
6. Kirsch, I. *et al.* 2008. Initial severity and antidepressant benefits: a meta-analysis of data submitted to the Food and Drug Administration. *PLoS Med.* **5:** e45.
7. Holmes, P.V. 2003. Rodent models of depression: reexamining validity without anthropomorphic inference. *Crit. Rev. Neurobiol.* **15:** 143–174.
8. Pryce, C.R. & E. Seifritz. 2011. A translational research framework for enhanced validity of mouse models of psychopathological states in depression. *Psychoneuroendocrinology* **36:** 308–329.
9. Woolf, S.H. 2008. The meaning of translational research and why it matters. *JAMA* **299:** 211–213.
10. Dinan, T. 2001. Novel approaches to the treatment of depression by modulating the hypothalamic–pituitary–adrenal axis. *Hum. Psychopharmacol.* **16:** 89–93.
11. Cowen, P.J. 2002. Cortisol, serotonin and depression: all stressed out? *Br. J. Psychiatry* **180:** 99–100.
12. Santarelli, L. *et al.* 2003. Requirement of hippocampal neurogenesis for the behavioral effects of antidepressants. *Science* **301:** 805–809.
13. Angelucci, F., S. Brene & A.A. Mathe. 2005. BDNF in schizophrenia, depression and corresponding animal models. *Mol. Psychiatry* **10:** 345–352.
14. Dantzer, R. 2009. Cytokine, sickness behavior, and depression. *Immunol. Allergy Clin. North Am.* **29:** 247–264.
15. Maes, M. 1993. A review on the acute phase response in major depression. *Rev. Neurosci.* **4:** 407–416.
16. Maes, M. 1995. Evidence for an immune response in major depression: a review and hypothesis. *Prog. Neuropsychopharmacol. Biol. Psychiatry* **19:** 11–38.
17. Schiepers, O.J., M.C. Wichers & M. Maes. 2005. Cytokines and major depression. *Prog. Neuropsychopharmacol. Biol. Psychiatry* **29:** 201–217.
18. Maes, M. *et al.* 1997. Acute phase proteins in schizophrenia, mania and major depression: modulation by psychotropic drugs. *Psychiatry Res.* **66:** 1–11.
19. Maes, M. *et al.* 1997. Lower serum zinc in major depression is a sensitive marker of treatment resistance and of the

immune/inflammatory response in that illness. *Biol. Psychiatry* **42**: 349–358.

20. Kent, S., R.M. Bluthe, K.W. Kelley & R. Dantzer. 1992. Sickness behavior as a new target for drug development. *Trends Pharmacol. Sci.* **13**: 24–28.

21. Hart, B.L. 1988. Biological basis of the behavior of sick animals. *Neurosci. Biobehav. Rev.* **12**: 123–137.

22. Dantzer, R. 2001. Cytokine-induced sickness behavior: mechanisms and implications. *Ann. N.Y. Acad. Sci.* **933**: 222–234.

23. Dantzer, R., J.C. O'Connor, G.G. Freund, *et al.* 2008. From inflammation to sickness and depression: when the immune system subjugates the brain. *Nat. Rev. Neurosci.* **9**: 46–56.

24. Yirmiya, R. *et al.* 2001. Effects of antidepressant drugs on the behavioral and physiological responses to lipopolysaccharide (LPS) in rodents. *Neuropsychopharmacology* **24**: 531–544.

25. Shen, Y., T.J. Connor, Y. Nolan, *et al.* 1999. Differential effect of chronic antidepressant treatments on lipopolysaccharide-induced depressive-like behavioural symptoms in the rat. *Life Sci.* **65**: 1773–1786.

26. Dredge, K., T.J. Connor, J.P. Kelly & B.E. Leonard. 1999. Differential effect of a single high dose of the tricyclic antidepressant imipramine on interleukin-1beta and tumor necrosis factor-alpha secretion following an in vivo lipopolysaccharide challenge in rats. *Int. J. Immunopharmacol.* **21**: 663–673.

27. Castanon, N., J.P. Konsman, C. Medina, *et al.* 2003. Chronic treatment with the antidepressant tianeptine attenuates lipopolysaccharide-induced Fos expression in the rat paraventricular nucleus and HPA axis activation. *Psychoneuroendocrinology* **28**: 19–34.

28. Dunn, A.J. & A.H. Swiergiel. 2001. The reductions in sweetened milk intake induced by interleukin-1 and endotoxin are not prevented by chronic antidepressant treatment. *Neuroimmunomodulation* **9**: 163–169.

29. Dunn, A.J. & A.H. Swiergiel. 2005. Effects of interleukin-1 and endotoxin in the forced swim and tail suspension tests in mice. *Pharmacol. Biochem. Behav.* **81**: 688–693.

30. Dunn, A.J., A.H. Swiergiel & R. de Beaurepaire. 2005. Cytokines as mediators of depression: what can we learn from animal studies? *Neurosci. Biobehav. Rev.* **29**: 891–909.

31. Petit-Demouliere, B., F. Chenu & M. Bourin. 2005. Forced swimming test in mice: a review of antidepressant activity. *Psychopharmacology* **177**: 245–255.

32. Costa-Pinto, F.A., D.W. Cohn, V.M. Sa-Rocha, *et al.* 2009. Behavior: a relevant tool for brain-immune system interaction studies. *Ann. N.Y. Acad. Sci.* **1153**: 107–119.

33. Cohn, D.W., D. Kinoshita & L.C. Sá-Rocha. 2012. *Lipopolysaccharide administration in the dominant mouse destabilizes social hierarchy.* *Behav. Processes.* DOI:10.1016/j.beproc.2012.05.008

34. Cohn, D.W.H. & L.C. de Sa-Rocha. 2009. Sickness and aggressive behavior in dominant and subordinate mice. *Ethology* **115**: 112–121.

35. Gomes, K.S. *et al.* 2009. Contrasting effects of acute and chronic treatment with imipramine and fluoxetine on inhibitory avoidance and escape responses in mice exposed to the elevated T-maze. *Brain Res. Bull.* **78**: 323–327.

36. Mason, S.S. *et al.* 2009. Differential sensitivity to SSRI and tricyclic antidepressants in juvenile and adult mice of three strains. *Eur. J. Pharmacol.* **602**: 306–315.

37. Gray, S., & J.L. Hurst. 1995. The effects of cage cleaning on aggression within groups of male laboratory mice. *Anim. Behav* **49**: 821–826.

38. Palanza, P., L. Gioiosa & S. Parmigiani. 2001. Social stress in mice: gender differences and effects of estrous cycle and social dominance. *Physiol. Behav.* **73**: 411–420.

39. Slattery, D.A., A.L. Hudson & D.J. Nutt. 2004. Invited review: the evolution of antidepressant mechanisms. *Fundam. Clin. Pharmacol.* **18**: 1–21.

40. Aubert, A. 1999. Sickness and behaviour in animals: a motivational perspective. *Neurosci. Biobehav. Rev.* **23**: 1029–1036.

Ann. N.Y. Acad. Sci. ISSN 0077-8923

ANNALS OF THE NEW YORK ACADEMY OF SCIENCES
Issue: *Neuroimmunomodulation in Health and Disease*

Peripheral immune system and neuroimmune communication impairment in a mouse model of Alzheimer's disease

Lydia Giménez-Llort,[1] Ianire Maté,[2] Rashed Manassra,[2] Carmen Vida,[2] and Mónica De la Fuente[2]

[1]Department of Psychiatry and Forensic Medicine, Institute of Neurosciences, Universitat Autònoma de Barcelona, Bellaterra, Spain. [2]Department of Physiology, Faculty of Biology, Complutense University of Madrid, Madrid, Spain.

Address for correspondence: Mónica De la Fuente, Department of Physiology (Animal Physiology II), Faculty of Biology, Complutense University of Madrid, 28040 Madrid, Spain. mondelaf@bio.ucm.es

Neurodegenerative diseases such as Alzheimer's disease (AD) can be understood in the context of the aging of neuroimmune communication. Although the contribution to AD of the immune cells present in the brain is accepted, the role of the peripheral immune system is less well known. The present review examines the behavior and the function and redox state of peripheral immune cells in a triple-transgenic mouse model (3 x Tg-AD). These animals develop both beta-amyloid plaques and neurofibrillary tangles with a temporal- and regional-specific profile that closely mimics their development in the human AD brain. We have observed age and sex-related changes in several aspects of behavior and immune cell functions, which demonstrate premature aging. Lifestyle strategies such as physical exercise and environmental enrichment can improve these aspects. We propose that the analysis of the function and redox state of peripheral immune cells can be a useful tool for measuring the progression of AD.

Keywords: aging; Alzheimer's disease; triple-transgenic for Alzheimer's disease (3xTg-AD) mice; immunosenescence; behavior; sex

Neuroimmunomodulation in aging

Currently, it is recognized that the three regulatory systems, namely the nervous, endocrine, and immune systems, are intimately linked and interdependent. Thus, it is accepted that a neuro-immune-endocrine system allows the preservation of homeostasis and therefore of health[1,2] (Fig. 1). The communication between these regulatory systems is mediated by cytokines, hormones, and neurotransmitters through the presence of their receptors on the cells of the three systems. Moreover, mediators of the three systems coexist in lymphoid, neural, and endocrine tissues. These facts show the complexity of the regulation not only at general levels, but also at local levels. This neuroendocrine–immune communication allows for the understanding of why depression, anxiety, and emotional stress are accompanied by a greater vulnerability to infections, cancers, and autoimmune diseases.[3–5]

In addition, the aging process, which may be defined as a progressive and general impairment of the functions of an organism that leads to a lower ability to adaptively react to changes and preserve homeostasis, affects all of the physiological systems but especially the regulatory systems and the communication between them (Fig. 1). The difficulty in preserving the homeostasis is the basis of the increase of age-related morbidity and mortality.[2,6,7] With aging, the nervous system suffers a progressive loss of function, which can be shown, for example, in such functions as sensation, cognition, memory, and motor activity. The hippocampus shows an age-related decrease of neurogenesis, which explains the learning and cognitive impairment in aged subjects.[8] In the endocrine system, there are also several changes that accompany healthy aging. These include, for example, the decrease of several hormones, such as the growth hormone and sexual hormones.[9] Moreover, the age-related disturbances of the

doi: 10.1111/j.1749-6632.2012.06639.x

Ann. N.Y. Acad. Sci. 1262 (2012) 74–84 © 2012 New York Academy of Sciences.

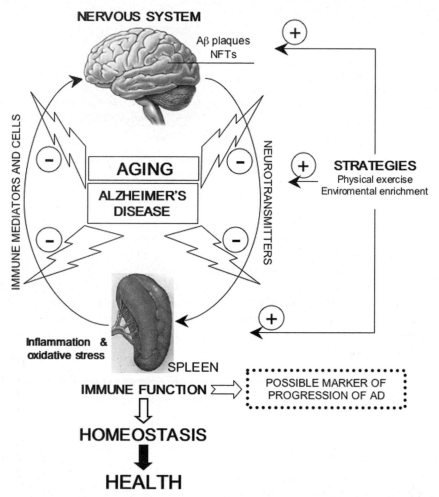

NERVOUS SYSTEM

Figure 1. With age, all regulatory systems (the nervous and immune system) involved in homeostasis show impairment due to an increase of inflammatory and oxidative stress. This age-related loss of homeostasis is exacerbated in AD and supports the hypothesis of a premature immunosenescence as a relevant factor of AD. Moreover, the analysis of immune function could be a marker of the progression of AD. Several strategies, such as physical exercise and environmental enrichment, seem to reestablish the homeostatic systems. Aβ plaques, beta amyloid plaques; NFTs, neurofibrillary tangles; (+), positive effect; (−), negative effect.

hypothalamic–pituitary–adrenal (HPA) axis are responsible for decreasing the stress adaptability in old subjects, this being, at least in part, the cause of their health impairment.[10] With respect to the immune system, it is presently accepted that its components undergo striking age-associated restructuring, leading to changes that may include enhanced as well as diminished functions. This is termed *immunosenescence*,[2,11,12] which appears with increasing age in both the periphery and in the central nervous system (CNS).

Age-related changes in the communication among the homeostatic systems were proposed to be the main cause of physiological senescence.[13] Although more than 300 hypotheses have been proposed to explain the aging process,[14] the free-radical hypothesis proposed by Harman[15] is now probably the most widely accepted explanation of how this process occurs. This view proposes that aging is the consequence of the accumulation of damage by deleterious oxidation in biomolecules caused by the high reactivity of free radicals produced in cells as a result of the necessary use of oxygen. Thus, the age-related changes in an organism are linked to chronic oxidative and inflammatory stress (a progressive imbalance between endogenous antioxidant/anti-inflammatory and oxidant/proinflammatory compounds, with higher levels of the latter), which affects all cells, especially those of the regulatory systems, thus explaining their impaired function.[2,6,7,12] Moreover, oxidation and inflammation are the basis of immunosenescence. In fact, although immune cells need to produce oxidative and inflammatory compounds in order to perform their defensive functions, when produced in relatively high concentrations these compounds cause damage to cells.[2,16] Since the functional capacity of the immune system has been considered the best marker of health,[17] the preservation of this functional and redox state is relevant to the maintenance of healthy and successful aging.[2] Moreover, we have proposed the concept of "oxidation inflammation in aging," characterized by a state of "oxi-inflamm-aging" in which age-related impairment of the immune system could affect the functions of the other regulatory systems through increased oxidative and inflammatory stress, thus resulting in the alteration of homeostasis and an increase in morbidity and mortality.[2,6,7,11]

In addition, aging is a very heterogeneous process; there are different rates of age-related physiological changes in each system or tissue of the organism and in the diverse members of a population of the same chronological age. This fact justifies the introduction of the concept of *biological age*, which determines the rate of aging experienced by each individual, and with a better predictive value for longevity than chronological age.[2,18] Biological age is related to mean longevity (the mean age at which members of a population that have been born on the same date live to); members of a population with a higher rate of aging have an older biological age and a shorter life span. If the maximum longevity (the maximum time that an individual belonging to a determined species can live) is fixed in each species, the mean life span of individual organisms shows marked variability and can be increased by environmental factors. In order to determine biological age, the parameters that change with age and are associated with premature death should be determined.[19] Since a positive relationship has been shown between good function of immune cells and longevity,[2,11,17,20] we have proposed several immune function parameters as markers of biological age and, therefore, predictors of longevity.[2] Moreover, the redox situation and inflammatory state of im-

mune cells are related to their functional capacity and the life span of an individual. Thus, the state of immune system allows us to know the rate of aging of the organism, and the immune system has the capacity to modify this rate.[2]

Neurodegenerative diseases in the context of the age-related changes in nervous and immune communication

Age-related neurodegenerative diseases can be considered in the context of nervous and immune system communication. Since the mediators of immune cells can affect the nervous system at peripheral and central levels it is necessary to consider both influences (Fig. 1). It is currently accepted that the immune cells present in the CNS have neuroprotective mechanisms. If they show uncontrolled responses, however, which is frequent in aging, they can be the cause of neurological disorders,[21,22] increasing an oxidation and inflammation situation.[8,15] Thus, the age-related, systemic, and progressive accumulation of immune cells in brain areas, such as the hippocampus, might have a causative impact on the progressive cognitive impairment that occurs with aging[23] and, in addition, may contribute to neuropathologies.[24] Age-related peripheral chronic inflammation can also influence cells in the CNS, leading to neuroinflammation and neuro-oxidation, which play a key role in the behavioral and cognitive deficits associated with aging and age-related diseases such as Alzheimer's disease (AD).[25] Thus, immunosenescence is important because the inflammatory signals that are initiated in the periphery can be propagated through CNS immune cells.[26] Interestingly, both humans and mice that achieve longevity under healthy conditions have been shown to preserve the functions and low peripheral inflammatory and oxidative status of the immune cells.[2,11,27]

Characteristics of AD

AD, the most common neurodegenerative disorder and cause of senile dementia, is a progressive age-related disease whose prevalence in the elderly is dramatically increasing with longer life expectancy and social aging. AD is characterized by deficits in memory, spatial vision, language, and executive function. While cognitive deficits have traditionally been emphasized in defining AD, there are a variety of neurobehavioral symptoms that are also commonly associated with the disease, including increased apathy, agitation, anxiety, and other psychiatric symptoms, such as delusions or hallucinations.[28] The main histopathological hallmarks of AD are extracellular amyloid plaques and neurofibrillary tangles (NFTs). The AD brain is further characterized by massive neuronal cell density and synapse number loss that appear to precede overt neuronal degeneration.[29] The most severe neuropathological changes occur in the hippocampus, followed by changes in the cortical and subcortical structures, including the amygdala.[30]

The component of the plaques is a 40–42 amino acid polypeptide beta-amyloid ($A\beta$, $A\beta_{40}$, and $A\beta_{42}$) that is derived from the larger amyloid precursor protein (APP) by proteolytic cleavage,[31] which is accomplished through the action of β-secretase and γ-secretase. This γ-secretase activity depends on a proteolytic complex[32,33] and dictates the length of $A\beta$–$A\beta_{42}$, being the more neurotoxic because of its propensity to readily aggregate into oligomers and fibrils.[34] The second histopathological hallmark of AD is the appearance of intraneuronal aggregates composed of the highly phosphorylated protein tau.[35] Under normal physiological conditions, tau stabilizes microtubules, but under pathological conditions, the basal phosphorylated ratio of tau increases,[36] which leads to its dissociation from microtubules and subsequent formation of intraneuronal aggregates that lead to neuronal dysfunction and synapses loss.[37] Although the amyloid hypothesis of AD has been the focus of the a majority of studies in this research area, it has been proposed that APP intracellular domain (AICD) levels, which are elevated in the brains of AD patients and cause hyperphosphorylation and aggregation of tau protein, can contribute to the pathology independent of $A\beta$.[38]

Besides the hallmark lesions of AD, other reactive processes occur such as inflammation[39] and oxidative stress.[40] A large body of evidence supports the hypothesis of a direct contribution of the inflammatory response to amyloid plaque progression, as well as hyperphosphorylation of tau protein, and thus to the neurodegeneration associated with AD. Moreover, $A\beta$ itself has been shown to act as a proinflammatory agent,[41] and many inflammatory mediators, such as cytokines, which are upregulated by $A\beta$, can serve to increase tau pathology.[42] Reactive oxygen species (ROS) are also produced as a result of this inflammatory response;[43] ROS can damage cells, which may further exacerbate this inflammatory state.

It is currently accepted that there is an associated immunological response in AD, but it is still unclear whether this is beneficial or harmful. In fact, several authors have proposed that in AD, the key pathogenic phenomena consist in the long-term maladaptive activation of innate immunity.[44,45] Many components of this immunity, which are expressed throughout the brain, are disregulated in AD and may act as a double-edged sword, with either beneficial or detrimental effects. Thus, the disturbed balance between complement activator and regulatory proteins seems to mediate neuronal lysis in AD.[46] Monocytes recruited into the brain exhibit ineffective $A\beta$ phagocytosis in AD patients.[47] Microglial cells (brain macrophages), which are found upregulated in the AD brain, attempt to phagocytose $A\beta$ plaques and secrete anti-inflammatory cytokines; they fail, however, at restricting $A\beta$ plaque formation, and their overactivation results in the secretion of proinflammatory cytokines, chemokines, and ROS, as well as hyperphosphorylations of tau protein, all of which exacerbate the pathology.[40,44,48] In addition, it has been suggested that $A\beta$ is an antimicrobial peptide that may normally function in the innate immune system.[49] Moreover, the presence in amyloid plaques and tangles of many immune-related proteins in addition to viral proteins has suggested that plaques and tangles represent cemeteries for a battle between the virus and the host's defense network.[50]

Murine models of AD

During the last two decades it has been a challenge to model behavioral and neuronal symptoms of AD.[51] Advances in gene transfer techniques and the identification of the genes implicated in the autosomal dominant familiar AD have made possible the development of many experimental models, particularly in mice. These target the major aspects of the neuropathological characteristics of AD, such as $A\beta$ plaques and NFTs. In this regard, several mutations in the genes of human APP, presenilin-1 (PS1), presenilin-2 (PS2), and protein tau have been reported to produce AD-like pathology. Nevertheless, a valid animal model for AD should exhibit progressive AD-like neuropathology

Table 1. Temporal course of behavioral changes observed in males and females of a Spanish colony of $3 \times$ Tg-AD versus NTg mice

Behavior parameters	Stages of neurodegeneration			
	Onset (2.5 m)	Early stages (4 m)	Moderate stages (6 m)	Advanced stages (12 m or more)
Increased sensorimotor function	n.s.	n.s.	+	++
BPSD-like symptoms				
Emotionality	+	+	++	+++
Neophobia	n.s.	+.	n.s.	+++
Reduced exploration in ansiogenic places	+	++	++	+++
Anxiety-like behaviors	+	++	++	+++
Hyperactivity	n.s.	+	+	++
Disinhibition	n.s.	n.a.	++	n.s.
Impulsivity	n.s.	n.a.	+	+++
Reduced novelty seeking	n.s.	n.a	n.s.	n.a.
Dysfunction of startle response	n.a.	n.a.	+	n.a.
Dysfunction of prepulse inhibition	n.a.	n.a.	+	n.a.
Cognition				
Spatial working memory deficits	n.s.	n.a.	+	+++
Spatial short-term memory deficits	n.s.	n.a.	n.s.	+++
Spatial long-term memory deficits	n.s.	+	++	+++
Instrumental conditioning deficits	n.a.	n.a.	++	n.a.
Altered circadian rhythms	n.a.	n.a.	+	+++

NOTE: Genotype effects: n.s., nonsignificant differences; n.a., not assessed in our colony of mice. +, ++, +++: higher levels in $3 \times$ Tg-AD than NTg mice ($P < 0.05$, $P < 0.01$, and $P < 0.001$, respectively). The appearance of AD pathology at different ages: 2.5 months (onset of AD; no neuropathological manifestation); 4 months (early stages; intracellular Aβ immunoreactivity); 6 months (moderate stages; extracellular Aβ deposits but still no tau alterations); 12 months (advanced stages; Aβ deposits in many cortical regions and tau hyperphosphorylation).[30,60,63–65]

and cognitive deficits that closely mimic human disease progression and should be verified in different laboratories.[52] Although several transgenic animal models of AD based on the expression of mutant familiar AD transgenes have been developed,[34] the triple-transgenic mouse ($3 \times$ Tg-AD, harboring APP$_{Swe}$ and tau$_{P301L}$ transgenes on a mutant PS1$_{M146V}$ knock-in background) represents a unique model that develops both Aβ plaques and NFTs with a temporal- and regional-specific profile that closely mimics their development in the human AD brain.[53,54] Thus, since LaFerla's laboratory created the triple-transgenic $3 \times$ Tg-AD mice in 2003,[53] the model has been the object of a considerable number of publications based on the singularity mentioned earlier. Moreover, this model manifests other hallmarks of the disease, such as the characteristic reactive gliosis inflammatory profile, choliner-

gic deficits, synaptic dysfunction, deficits in learning and memory, and the behavioral and psychological symptoms of dementia (BPSD) in an age-dependent manner.[50,53,55–57] The $3 \times$ Tg-AD mouse model has been used extensively to dissect pathogenic mechanisms and for therapeutic gene approaches[42] and is continually evolving, which should lead to a better transfer of mouse model–based therapies into the clinic.[34]

Age-related changes of the behavioral parameters in $3 \times$ Tg-AD mice

At the behavioral level, the $3 \times$ Tg-AD model provided evidence that the Aβ peptide causes the onset of early Alzheimer's disease–related cognitive deficits in learning and memory.[58] Thus, the earliest cognitive impairment manifests itself in these animals at 4 months as a deficit in long-term retention, which is correlated with accumulation of Aβ in the hippocampus and amygdala that worsens with aging and the advancement of the neuropathological stages.[51] In addition to the limbic-dependent spatial learning and memory deficits (observed in the Morris water maze), the 3×Tg-AD mice also show deficiencies in working memory[59] and in emotional learning using passive avoidance;[60] in addition, we have provided evidence of the influence of emotional behavior in fear-conditioned learning tasks.[61] Moreover, the 3×Tg-AD mice can also be considered a valuable animal model for AD because of its sex- and age-dependent course of severity of noncognitive disturbances resembling BPSD or neuropsychiatric symptoms.[62] These BPSD-like symptoms, such as increased anxiety-related behavior with reduction of exploratory activity, appear before the cognitive deficits, as early as 2.5 months of age, and are indicative of increased responsiveness to stressful situations.[51,62] In correlation with the appearance of intraneuronal Aβ immunoreactivity and further Aβ and tau pathologies, the 3×Tg-AD mice progressively exhibit increased neophobia, freezing behavior, reduced exploratory efficiency, and other behavioral variables indicative of reduced coping with stress.[51,60,62–65] In addition, there is a lack of regulation of behavior (i.e., impulsivity and disinhibition) and deficits in sensorimotor gating (in startle response and prepulse inhibition) that progressively worsens with the neuropathological effects of limbic areas[61] (Table 1).

The immune system in 3×Tg-AD mice—an early marker of the disease

The impairment in the neuroimmunendocrine network that occurs with aging is accelerated and more pronounced in the 3×Tg-AD mice.[65,67] With respect to the peripheral immune system, the organometrics of immune organs are relevant to the understanding of the physical variation and changes due to disease. Thus, thymus weight has been established as an indirect indicator of the immunological functional state. In 3×Tg-AD mice, we have reported that both total weight and relative weight of peripheral immunoendocrine organs, such as thymus, spleen, and adrenal glands, correlate with the sex-dependent impairment of the neuroimmunoendocrine network described at both initial[66] and advanced[65] stages of the AD neuropathology.

Table 2. Several functional parameters in immune cells from young to old 3×Tg-AD versus NTg female mice

Parameters (function)	3×Tg-AD vs. NTg			
	2.5 m	4 m	9 m	15 m
Chemotaxis	=	↓↓↓	↓↓↓	↑↑↑
Proliferation	=	↓↓	↓↓	=
NK	=	↓	=(↓)	=
IL-2 secretion	n.a.	↓	=(↓)	↑

NOTE: The appearance of AD pathology at different ages: 2.5 months (onset of AD; no neuropathological manifestation); 4 months (early stages; intracellular Aβ immunoreactivity); 9 months (moderate stages; extracellular Aβ deposits but still no tau alterations); and 15 months (advanced stages; Aβ deposits in many cortical regions and tau hyperphosphorylation). ↓, ↓↓, and ↓↓↓: lower levels in 3×Tg-AD than NTg mice (P < 0.05, P < 0.01, and P < 0.001, respectively); ↑ and ↑↑↑: higher levels in 3×Tg-AD than NTg mice (P < 0.05 and P < 0.001, respectively); =: similar levels in 3×Tg-AD and NTg mice; =(↓): decreased tendency without significant differences in 3×Tg-AD with respect to NTg mice; n.a., not assessed in our colony of mice. Refs. 65, 67–70.

In relation to the functional parameters of the immune cells we have analyzed, several have been proposed as biological age markers, such as chemotaxis, proliferation in response to mitogens (ConA and LPS), antitumoral NK activity, and IL-2 secretion.[2,6,7,12] We have observed an age-related decrease in these functions and that subjects with lower values of these parameters show an older biological age and earlier death than those with the same chronological age that have higher values.[2,7,12] Moreover, individuals reaching a high longevity (human centenarians and long-lived mice) show these functions with values similar to those in healthy adults.[2,11,27] In 3×Tg-AD mice, we have studied these parameters in immune cells from peritoneum, spleen, and thymus, and the changes in comparison to NTg animals. The changes observed were similar in cells from all these locations (Table 2).[65,68] As can be seen in 3×Tg-AD mice, there is a decrease in these functions with respect to the NTg animals in young adult and adult mice (4 and 9 months of age, respectively). In young mice (2.5 months of age) the differences have not appeared yet, and older (15 months of age) 3×Tg-AD mice show an increase in chemotaxis and IL-2 secretion.[65,68–70] With respect to the oxidative stress parameters analyzed (Table 3), both young adult and adult transgenic mice, compared with NTg animals, show a decrease in spleen antioxidant defenses, such as total glutathione (TG) levels and activity of GPx and GR antioxidant enzymes as well as an increase in xantin oxidase (XO) activity (an enzyme that produces oxidant compounds).[72,73] These changes in antioxidants and oxidant compounds are characteristics of prematurely and chronologically aged subjects.[2,12,16,74] Thus, there is premature immunosenescence in the 3×Tg-AD mice, which is underpinned by an oxidative stress state. These facts confirm the premature aging of transgenic mice and explain their early mortality. Moreover, the analysis of the peripheral immune functions seems to be a good marker of the progression of AD. All this confirms the idea of the early involvement of immunity in the pathogenesis of AD.[45]

Strategies to improve behavior and the immune system in 3×Tg-AD mice

We have proposed several lifestyle strategies, such as the performance of physical exercise and environmental enrichment (EE), to improve the redox state, immune function, and behavioral response in aging mice and in several animal models proposed as premature aging models.[2,7,12] These strategies retard the aging process, improving homeostasis and increasing the longevity of the individuals.[7,12] Thus, these strategies could also exert beneficial effects on 3×Tg-AD mice, which show high oxidative stress, impaired peripheral immune cell functions, and shorter longevity.[64,65,67]

Physical exercise in 3×Tg-AD mice

It is well known that physical exercise is an effective means of preventing or delaying chronic diseases. Together with the muscle and cardiovascular systems, physical activity strongly modulates the regulatory systems, providing beneficial effects in behavioral response and immune functions. Moreover, physical exercise is a physiological stress model, highlighting the relevance of interactions between the regulatory systems.[12,75]

In 3×Tg-AD mice, we have studied the effects of forced and voluntary exercise in both

Table 3. Oxidative stress parameters (antioxidants and oxidants) in the spleen of young and adult 3×Tg-AD versus NTg mice

Parameters	3×Tg-AD vs. NTg
Antioxidant	
TG	↓↓↓
GPx	↓↓
GR	↓
Oxidant	
XO	↑↑

NOTE: ↓, ↑↑, and ↓↓↓: lower levels and activity in 3×Tg-AD than NTg mice (P < 0.05, P < 0.01, and P < 0.001, respectively); ↓↓: higher activity in 3×Tg-AD than NTg mice (P < 0.01).[64,66,71–73]

male and female 3×Tg-AD mice compared with nontransgenic mice. An exhaustive treadmill exercise administered at a moderate stage of AD neurodegeneration (7 months of age) partially protected 3×Tg-AD mice both in brain and peripheral organ functions, whereas voluntary physical exercise in a freely available running wheel ameliorated many of the 3×Tg-AD pathological behaviors, brain oxidative stress changes, as well as immune function.[64,66] Thus, exercise exerts an influence in addition to the improvement of physical condition in 3×Tg-AD mice.[64]

Environmental enrichment

EE is a good experimental model for the maintenance of a life that is socially, mentally, and physically active. The most common EE protocol in rodents is grouped housing using large cages with a variety of objects, which are changed frequently. This more complex and stimulating habitat, as opposed to the regular monotonous housing, induces sensory, cognitive, motor, and social stimulation.[76] EE reverses many of the adverse effects of the aging process on behavior, immune function, and oxidative stress, extending the life span of animals.[77,78] It is also known that EE reduces brain pathology and improves cognition and behavioral responses in a variety of murine models for age-related neurodegenerative diseases, including AD.[79] We have shown that EE can benefit several immune functions in old male 3×Tg-AD mice.[67]

Table 4. Gender differences in several parameters of function and redox state in immune cells from young, adult, and old NTg, and 3×Tg-AD mice

	Female vs. male					
	4 m		9 m		15 m	
Parameters	NTg	3× Tg-AD	NTg	3× Tg-AD	NTg	3× Tg-AD
Function						
Chemotaxis	<*	>***	>***	>*	>	>***
Proliferation						
Basal	<***	=	>*	>**	=	=
ConA	<***	<***	<***	=	=	>
LPS	<***	<***	<***	>*	=	>**
NK	<**	<**	>**	>**	=	=
Antioxidant						
TG	>**	=	>*	>**	>*	>*
GPx	=	=	=	=	n.a.	n.a.
GR	=	=	=	=	n.a.	n.a.
Oxidant						
XO	=	=	=	<	n.a.	n.a.

NOTE: "<, <*, <**, and <***", lower levels and activities in female versus male NTg and 3×Tg-AD mice (not significant (n.s.), $P < 0.05$, $P < 0.01$, and $P < 0.001$, respectively); >, >*, >**, and >***, higher levels and activities in female versus male NTg and 3×Tg-AD mice (n.s., $P < 0.05$, $P < 0.01$, and $P < 0.001$, respectively); =, similar levels and activities in female and male NTg and 3×Tg-AD; n.a., not assessed in our colony of mice.[65,67,69]

Sex differences in behavior and the immune system in 3×Tg-AD mice and physical exercise and environmental enrichment

There are sex differences in the changes of the neuroimmunoendocrine network with aging, and it is known that females live longer than males in a great range of animal species.[79] In general, females show stronger immune responses than males.[80,81] Although the prevalence of AD is higher in females, the higher vulnerability of the neuroimmunoendocrine network in males could result in a higher susceptibility to the deleterious effects of aging and could be responsible for the increased morbidity and mortality observed in 3×Tg-AD male mice.[65–67] This is possibly a consequence of the regulatory role of sex hormones promoting AD pathogene-

sis.[58] There is a sex-dependent behavioral phenotype that is exacerbated at early and moderate stages of the disease but loses relevance with the advancement of the neuropathology.[51,53,55,60,62,65] In general, at moderate pathological stages of the disease, male 3×Tg-AD mice show more homeostasis redox derangement that females, while females show greater brain AD pathology compared with males.[64] There are sex-related differences in functions and redox state of immune cells from 3×Tg-AD mice, which are in some cases different than those found in NTg mice, and depend on the age of animals (Table 4). The 3×Tg-AD male mice show worse immune functions and antioxidant levels than females, especially adult and old animals.[65,70] Thus, transgenic male mice show premature immunosenescence and aging with respect to females, starting at adult age.

There were sex differences in the effects of the previously mentioned strategies, both exercise[64,66] and EE,[67] on the behavior and immune functions of 3×Tg-AD, with males showing better responses than females.

Conclusions and proposals

Our results in 3×Tg-AD mice show the premature peripheral immunosenescence of these animals and seem to confirm the involvement of systemic immunity and inflammation in behavioral and cognitive deficits, such as those in AD,[82] as well as in the acceleration of the progression of the disease.[83] For this, we propose the study of the peripheral immune system as a marker of the course of AD and the effects of preventive and/or therapeutic interventions. Moreover, interaction effects among age, sex, genotype, and treatment should always be taken into consideration when assessing the outcome of those interventions.

Acknowledgments

We thank Mr. D. Potter for his help with the English language revision of the manuscript. This work was supported by the Spanish Ministry of Science and Innovation (BFU2008-04336, BFU2011–30336), ISC3 PI10/002831, Research Group of UCM (910379) and the Ministry of Health (RETICEF, RD06/0013/003). I. Mate has a fellowship from the "Gobierno Vasco" (BFI09-52).

Conflicts of interest

The authors declare no conflicts of interest.

References

1. Besedovsky, H. & A. Del Rey. 2007. Physiology of psychoneuroimmunology: a personal view. *Brain Behav. Immun.* **21:** 34–44.

2. De la Fuente, M. & J. Miquel. 2009. An update of the oxidation-Inflammation theory of aging: the involvement of the immune system in oxi-inflamm-aging. *Curr. Pharm. Des.* **15:** 3003–3026.

3. Arranz, L. *et al.* 2007. Impairment of several immune functions in anxious women. *J. Psychosom. Res.* **62:** 1–8.

4. Arranz, L. *et al.* 2009. Impairment of immune function in the social excluded homeless population. *Neuroimmunomodulation* **16:** 251–260.

5. Costa-Pinto, F.A. & J. Palermo-Nieto. 2010. Neuroimmune interactions in stress. *Neuroimmunomodulation* **17:** 196–199.

6. De la Fuente, M. 2008. Role of neuroimmunomodulation in aging. *Neuroimmunomodulation* **15:** 213–223.

7. De la Fuente, M. & L. Gimenez-Llort. 2010. Models of aging of neuroimmunomodulation: strategies for its improvement. *Neuroimmunomodulation* **17:** 213–216.

8. Couillard-Depres, S., B. Iglseder & L. Aigner. 2011. Neurogenesis, cellular plasticity and cognition: the impact of stem cells in the adult and aging brain. *Gerontology* **57:** 559–564.

9. Makrantonaki, E. *et al.* 2010. Skin and brain age together: the role of hormones in the ageing process. *Exp. Gerontol.* **45:** 801–813.

10. Lupien, S.J. *et al.* 2009. Effects of stress throughout the lifespan on the brain, behaviour and cognition. *Nature Rev. Neurosci.* **10:** 434–445.

11. Alonso-Fernández, P. & M. De la Fuente. 2011. Role of the immune sytem in aging and longevity. *Curr. Aging Sci.* **4:** 78–100.

12. De la Fuente, M. *et al.* 2011. Strategies to improve the functions and redox state of the immune system in aged subjects. *Curr. Pharm. Dise,* **17:** 3966–3993.

13. Fabris, N. 1990. A neuroendocrine-immune theory of aging. *Int. J. Neurosci.* **51:** 373–375.

14. Medvedev, Z.A. 1990. An attempt at a rational classification of theories of aging. *Biol. Rev.* **65:** 375–398.

15. Harman, D. 1956. Aging: a theory based on free radical and radiation chemistry. *J. Gerontol.* **2:** 298–300.

16. De la Fuente, M. *et al.* 2005. The immune system in the oxidation stress conditions of aging and hypertension. Favorable effects of antioxidants and physical exercise. *Antioxid. Redox Signal* **7:** 1356–1366.

17. Wayne, S.L. *et al.* 1990. Cell-mediated immunity as a predictor of morbidity and mortality in subjects over 60. *J. Gerontol.* **45:** M45–M48.

18. Borkan, G.A. & A.H. Norris. 1980. Assessment of biological age using a profile of physical parameters. *J. Gerontol.* **35:** 177–184.

19. Bulpitt, C.J. *et al.* 2009. Mortality according to a prior assessment of biological age. *Curr. Aging Sci.* **2:** 193–199.

20. Guayerbas, N. *et al.* 2002. Leukocyte function and life span in a murine model of premature immunosenescence. *Exp. Gerontol.* **37:** 249–256.

21. Gemma, C. 2010. Neuroimmunomodulation and aging. *Aging Dis.* **1:** 169–172.

22. Pizza, V. *et al.* 2011. Neuroinflamm-aging and neurodegenerative diseases: an overview. *CNS Neurol. Disord. Drug Targets* **10:** 621–634.

23. Stichel, C.C. & H. Luebbert. 2007. Inflammatory processes in the aging mouse brain: participation of dendritic cells and T-cells. *Neurobiol. Aging* **28:** 1507–1521.

24. Marx, F. *et al.* 1998. The possible role of the immune system in Alzheimer's disease. *Exp. Gerontol.* **33:** 871–881.

25. Von Bernhardi, R. 2007. Glial cell dysregulation: a new perspective on Alzheimer disease. *Neurotox. Res.* **12:** 215–232.

26. Corona, A.W. *et al.* 2012. Cognitive and behavioural consequences of impaired immunoregulation in aging. *J. Neuroimmune Pharmacol.* **7:** 7–23.

27. Arranz, L. *et al.* 2010. Preserved immune functions and controlled leukocyte oxidative stress in naturally long-lived mice: possible role of nuclear factor-kappa β. *J. Gerontol. A. Biol. Sci. Med. Sci.* **65:** 941–950.

28. Assal, F. & J.L. Cummings. 2002. Neuropsychiatric symptoms in the dementias. *Curr. Opin. Neurol.* **15:** 445–450.

29. Selkoe, D.J. 2002. Alzheimer's disease is a synaptic failure. *Science.* **298:** 1102–1111.

30. Arnold, S.E. *et al.* 1991. The topographical and neuroanatomical distribution of neurofibrillary tangles and neuritic plaques in the cerebral cortex of patients with Alzheimer's disease. *Cereb. Cortex* **1:** 103–116.

31. Glenner, G.G & C.W. Wong. 1984. Alzheimer's disease: initial report of the purification and characterization of a novel cerebrovascular amyloid protein. *Biochem. Biophys. Res. Commun.* **120:** 885–890.

32. Götz, J. *et al.* 2011. Models of Aβ toxicity in Alzheimer's disease. *Cell Mol. Life Sci.* **68:** 3359–3375.

33. Edbauer, D. 2003. Reconstitution of gamma-secretase activity. *Nat. Cell Biol.* **5:** 486–488.

34. Morrissette, D.A. *et al.* 2009. Relevants of transgenic mouse model to human Alzheimer disease. *J. Biol. Chem.* **284:** 6033–6037.

35. Grundke-Iqbal, I. *et al.* 1986. Abnormal phosphorylation of the microtubule-associated protein tau in Alzheimer cytoskeletal pathology. *Proc. Natl. Acad. Sci. U.S.A.* **83:** 4913–4917.

36. Kopke, E. *et al.* 1993. Microtubule-associated protein tau abnormal phosphorylation of a non-paired helical filament pool in Alzheimer disease. *J. Biol. Chem.* **268:** 24374–24384.

37. Iqbal, K. *et al.* 2005. Tau pathology in Alzheimer disease and other tauopathies. *Biochim. Biophys. Acta.* **1739:** 198–210.

38. Ghosal, K. *et al.* 2009. Alzheimer's disease-like pathological features in transgenic mice expressing the APP intracellular domain. *Proc. Natl. Acad. Sci.* **106:** 18367–18372.

39. Eikelenboom, P. *et al.* 2006. The significance of neuroinflammation in understanding Alzheimer's disease. *J. Neural. Transm.* **113:** 1685–1695.

40. Akiyama, H. *et al.* 2000. Inflammation and Alzheimer's disease. *Neurobiol. Aging* **21:** 383–421.

41. Matsuoka, Y. *et al.* 2001. Inflammatory responses to amyloidosis in a transgenic mouse model of Alzheimer's disease. *Am. J. Pathol.* **158:** 1345–1354.

42. Blurton-Jones, M. & F.M. Laferla. 2006. Pathways by which Aβ facilitates tau pathology. *Curr. Alzheimer Res.* **3:** 437–448.

43. Steele, M., G. Stuchbury & G. Munch. 2007. The molecular basis of the prevention of Alzheimer's disease through healthy nutrition. *Exp. Gerontol.* **42:** 28–36.

44. Maccioni, R.B. *et al.* 2009. The role of neuroimmunomodulation in Alzheimer's disease. *Ann. N.Y. Acad. Sci.* **1153:** 240–246.

45. Eikelenboom, P. *et al.* 2011. The early involvement of the innate immunity in the pathogenesis of late-inset Alzheimer's disease: neuropathological, epidemiological and genetic evidence. *Curr. Alzheimer Res.* **8:** 142–150.

46. Veerhuis, R. *et al.* 2011. Complement in the brain. *Mol. Immunol.* **48:** 1592–1603.

47. Feng, Y. *et al.* 2011. Monocytes and Alzheimer's disease. *Neurosci. Bull.* **27:** 115–122.

48. Solito, E. & M. Sastre. 2012. Microglia function in Alzheimer's disease. *Frontier Pharmacol.* doi:10.3389/phar.2012.00014.

49. Soscia, A. *et al.* 2010. The Alzheimer's disease-associated amyloid β-protein is an antimicrobial peptide. *PLoS ONE* **5:** e9505.

50. Carter, C.J. 2011. Alzheimer's disease plaques and tangles: cementeries of a pyrrhic victory of the immune defense network against herpes simplex infection at the exposure of complement and inflammation-mediated neuronal destruction. *Neurochem. Int.* **58:** 301–320.

51. Giménez-Llort, L. *et al.* 2007. Modeling behavioral and neuronal symptoms of Alzheimer's disease in mice: a role for interpersonal amyloid. *Neurosci. Biobehav. Rev.* **31:** 125–147.

52. Janus, C. & D. Westaway. 2001. Transgenic mouse models of Alzheimer's disease. *Physiol. Behav.* **73:** 873–886.

53. Oddo, S. *et al.* 2003. Triple-transgenic model of Alzheimer's disease with plaques and tangles intracellular Aβ and synaptic disfunction. *Neuron.* **39:** 409–421.

54. Mastrangelo, M.A. & W.J. Bowers. 2008. Detailed immunohistochemical characterization of temporal and spatial progression of Alzheimer's disease related pathologies in male triple-transgenic mice. *BCM Neurosci.* **9:** 81.

55. Oddo, S. *et al.* 2003. Amyloid deposition precedes tangles formation in a triple model of Alzheimer's disease. *Neurobiol. Aging* **24:** 1063–1070.

56. Oddo, S. *et al.* 2005. Chronic nicotine administration exacerbates tau pathology in transgenic models of Alzheimer's disease. *Proc. Natl. Acad. Sci. U.S.A.* **102:** 3046–3051.

57. Kitazawa, M. *et al.* 2005. Lipopolisaccharide-induced inflammation exacerbates tau pathology by a cycling-dependent kinase 5-mediated pathway in a transgenic model of Alzheimer's disease. *J. Neurosci.* **25:** 8843–8853.

58. Billings, L. *et al.* 2005. Intraneuronal Aβ causes the onset of early Alzheimer's disease-related cognitive deficits in transgenic mice. *Neurone* **45:** 675–688.

59. Rosario, E.R. *et al.* 2006. Androgens regulate the development of neuropathology in a triple-transgenic mouse model of Alzheimer's disease. *J. Neurosci.* **26:** 13384–13389.

60. Clinton, L.K. *et al.* 2007. Age-dependent sexual dimorphism in cognition and stress response in 3xTg-AD mice. *Neurobiol. Dis.* **28:** 76–82.

61. España, J. *et al.* 2010. Intraneuronal β-amyloid accumulation in the amygdala enhances fear and anxiety in Alzheimer's disease transgenic mice. *Biol. Psychiatry* **67:** 513–521.

62. Giménez-Llort, L. *et al.* 2006. Modeling neuropsychiatric symptoms of Alzheimer's disease dementia in 3xTg-AD mice. In *Alzheimer's Disease: New Advances.* K. Iqbal, B. Winblad, and J. Avila, Eds.: 513–516. Medimond, Englewood, NJ. USA.

63. Sterniczuk, R. *et al.* 2010. Characterization of the 3xTg-AD mouse model of Alzheimer's disease: part 2. Behavioral and cognitive changes. *Brain Res.* **1348:** 149–155.

64. García-Mesa, Y. *et al.* 2011. Physical exercise protects against Alzheimer's disease in 3xTg-AD mice. *J. Alz. Dis.* **23:** 1–34.

65. Giménez-Llort, L. *et al.* 2008. Gender-specific neuroimmunoendocrine aging in a triple-transgenic 3xTg-AD mouse model for Alzheimer's disease and its relation with longevity. *Neuroimmunomodulation* **15:** 331–343.

66. Giménez-Llort, L. *et al.* 2010. Genger-specific neuroimmunoendocrine response to treadmill exercise in 3xTg-AD mice. *Int. J. Alzheimers Dis.* doi:10.4061/2010/128354.

67. Arranz, L. *et al.* 2011. Effect of environmental enrichment on the immunoendocrine aging of male and female triple-transgenic 3xTg-AD mice for Alzheimer's disease. *J. Alz. Dis.* **4:** 727–737.

68. Sainz de Aja, J.G. *et al.* 2011. Differences in several functions and catalase activity in peritoneal leukocytes of young and mature triple-transgenic mice for Alzheimer's disease. *Neuroimmunomodulation* **18:** 403.

69. De la Fuente, M. *et al.* 2011. Impairment of several behavioral parameters and spleen leukocyte functions in male and female triple transgenic mice for Alzheimer's disease (3xTgAD) at the early age of 4 months. *Neuroimmunomodulation* **18:** 371.

70. De la Fuente, M. *et al.* 2011. The impairment of the lymphocyte functions in female triple transgenic for Alzheimer's disease (3xTg-AD) mice is not increased by ovariectomy. *Neuroimmunomodulation* **18:** 371.

71. Hernanz, A. *et al.* 2009. Increase of xanthine oxidase activity in several organs from triple-transgenic adult mice for Alzheimer's disease. *J. Nutr. Health Aging* **13:** S553.

72. Giménez-Llort, L. *et al.* 2009. Decrease of the antioxidant defenses in 3xTg-AD triple-transgenic adult mice for Alzheimer's disease. *Alzheimer's Dement.* **5:** P502.

73. Vida, C. *et al.* 2011. The age-related increase in xanthine oxidase expression and activity in several tissues from mice is not shown in long-lived animals. *Biogerontology* **12:** 551–564.

74. Fragala, M.S. *et al.* 2011. Neuroendocrine-immune interactions and responses to exercise. *Sports Med.* **41:** 621–639.

75. Nithianantharajan, J. & A.J. Hannan. 2006. Enriched environments, experience-dependent plasticity and disorders of nervous system. *Nat. Rev. Neurosci.* **7:** 697–709.

76. Zambrana, C. *et al.* 2007. Influence of aging and enriched environment on motor activity and emotional responses in mice. *Ann. N.Y. Acad. Sci.* **1100:** 543–552.

77. Arranz, L. *et al.* 2010. Environmental enrichment improves age-related immune system impairment. Long-term exposure since adulthood increases life span in mice. *Rejuvenation Res.* **13:** 415–428.

78. Görtz, N. *et al.* 2008. Effects of environmental enrichment on exploration, anxiety and memory in female TgCRND8 Alzheimer mice. *Behav. Brain Res.* **191:** 43–48.

79. Rosenblitt, J.C. *et al.* 2001. Sensation seeking and hormones in men and women: exploring the link. *Horm. Behav.* **40:** 396–402.

80. Aspinall, R. 2000. Longevity and the immune response. *Biogerontology* **1:** 273–278.

81. De la Fuente, M. *et al.* 2004. Changes with aging in several leukocyte functions of male and female rats. *Biogerontology* **5:** 389–400.

82. Pellicano, M. *et al.* 2010. Systemic immune responses in Alzheimer's disease: in vitro mononuclear cell activation and cytokine production. *J. Alz. Dis.* **21:** 181–192.

83. Perry, V.H. 2010. Contribution of systemic inflammation to chronic neurodegeneration. *Acta Neuropathol.* **120:** 277–286.

Ann. N.Y. Acad. Sci. ISSN 0077-8923

ANNALS OF THE NEW YORK ACADEMY OF SCIENCES
Issue: *Neuroimmunomodulation in Health and Disease*

Effects of plasmalogens on systemic lipopolysaccharide-induced glial activation and β-amyloid accumulation in adult mice

Toshihiko Katafuchi,[1] Masataka Ifuku,[1] Shiro Mawatari,[2] Mami Noda,[3] Kiyotaka Miake,[4] Masaaki Sugiyama,[4] and Takehiko Fujino[2]

[1]Department of Integrative Physiology, Graduate School of Medical Sciences, Kyushu University, Fukuoka, Japan. [2]Institute of Rheological Function of Food, Kasuya-gun, Fukuoka, Japan. [3]Laboratory of Pathophysiology, Graduate School of Pharmaceutical Sciences, Kyushu University, Fukuoka, Japan. [4]Central Research Institute, Marudai Food Co. Ltd., Osaka, Japan

Address for correspondence: Toshihiko Katafuchi, M.D., Ph.D., Department of Integrative Physiology, Graduate School of Medical Sciences, Kyushu University, Fukuoka 812-8582, Japan. kataf@physiol.med.kyushu-u.ac.jp

Neuroinflammation essentially involves an activation of glial cells as the cause/effect of neurodegenerative diseases such as Alzheimer's disease (AD). Plasmalogens (Pls) are glycerophospholipids constituting cellular membranes and play significant roles in membrane fluidity and cellular processes like vesicular fusion and signal transduction. Intraperitoneal (i.p.) injection of lipopolysaccharide (LPS, 250 μg/kg) for 7 days resulted in the morphological changes and increase in number of Iba-1$^+$ microglia showing neuroinflammation in the adult mouse hippocampus. The LPS-induced activation of glial cells was significantly attenuated by i.p. pretreatment with Pls dissolved in corn oil. In addition, systemic injection of LPS induced $A\beta_{1-16}{}^+$ neurons in the hippocampus were also abolished by application of Pls. Finally, contents of Pls in the hippocampus decreased after LPS injection, and the reduction was suppressed by administration of Pls. These findings suggest an antiamyloidogenic effect of Pls, implicating a possible therapeutic application of Pls against AD.

Keywords: neuroinflammation; phospholipids; microglia; Alzheimer's disease

Introduction

It has been shown that systemic administration of lipopolysaccharide (LPS), a ligand for Toll-like receptor (TLR) 4, induces an impairment of cognitive behavior in mice.[1–3] Lee *et al.*[2] have reported that the production of β-amyloid protein (Aβ) and the activity of β-secretase, a key rate-limiting enzyme that initiates Aβ formation in the cortex and hippocampus increased in adult, but not young mice, following a single intraperitoneal (i.p.) injection of LPS. Furthermore, an intracellular accumulation of Aβ in the pyramidal neurons of the hippocampus was immunohistochemically demonstrated after daily LPS injections for 7 days.[2] Although the precise mechanisms for the LPS-induced amyloidogenesis have not yet been determined, it is likely that

neuroinflammation, which is characterized by activation of glial cells and increased expression of cytokines, chemokines, and reactive oxygen/nitrogen species (ROS/RNS), plays significant roles, since glial cell–mediated inflammatory responses induce Aβ deposition in the brain.[4–6]

Plasmalogens (Pls) are unique glycerophospholipids that contain a vinyl ether bond at the *sn*-1 position of the glycerol moiety. They are found in all mammalian tissues, especially in heart and brain as ethanolamine Pls (PlsEtn), which are much more abundant than choline Pls (PlsCho).[7] Pls release either docosahexaenoic acid (DHA) or arachidonic acid (ARA) from the *sn*-2 position through the activation of Pls-selective phospholipase A_2 (Pls-PLA$_2$).[8,9] Pls are not only structural membrane components and reservoirs for second messengers,

doi: 10.1111/j.1749-6632.2012.06641.x

but they are also involved in membrane fusion, ion transport, and cholesterol efflux.[7] In addition, the vinyl ether bond at the *sn*-1 position makes Pls more susceptible to oxidative stress than the corresponding ester-bonded glycerophospholipids, thereby acting as antioxidants and protecting cells from oxidative stress.[10–13]

It has been shown that patients suffering from Alzheimer's disease (AD) have reduced PlsEtn levels in the cortex and hippocampus.[14–16] The reduction of PlsEtn seems to be specific since other neurodegenerative diseases, such as Huntington's and Parkinson's disease, do not show the decrease in the corresponding affected brain regions (caudate nucleus and substantia nigra, respectively).[7,14,17] Furthermore, circulating PlsEtn levels are also decreased depending on the severity of dementia.[18,19] Therefore, it is possible that Pls are involved in the pathology of AD. In this study, we focused on the LPS-induced amyloidogenesis and sought to elucidate (1) the effects of plasmalogen on neuroinflammation and Aβ accumulation in the hippocampus, and (2) changes in hippocampal plasmalogen content following peripheral administration of LPS in adult mice.

Materials and methods

All of the experimental procedures involving the use of animals were approved by the Ethics Committee on Animal Experiments, Kyushu University, and were in accordance with the Guiding Principles for the Care and Use of Animals of the Physiological Society of Japan. All efforts were made to minimize animal suffering and the number of animals used for the studies.

Animals
Male C57/6J mice weighing 25–30 g (10 months old) were used in all experiments. Animals were housed in three cages (five per cage) at a temperature of 22 ± 2 °C with 12 h light/12 h dark cycle (light on at 8:00) and had free access to laboratory food and water. The mice were randomly divided into three groups: control, LPS, and LPS + Pls groups. LPS (Sigma-Aldrich, St. Louis, MO) was dissolved in saline, while Pls were in corn oil and then sonicated to ensure complete solubilization. The LPS group received intraperitoneal (i.p.) injection of LPS (250 μg/kg) followed by vehicle for Pls, corn oil, in the morning (9:00–10:00) daily for 7 days before sacrifice. The LPS + Pls group was treated with LPS and Pls (20 mg/kg), while the control group was given saline and corn oil for 7 days.

Plasmalogen preparation
Pls used in this study were prepared from chicken breast muscle by the same method as reported previously.[20] A high-performance liquid chromatography (HPLC) for phospholipids separation[21] indicated that the purified Pls consisted of 47.6% PlsEtn, 49.3% PlsCho, 2.4% sphingomyelin (SM), and 0.5% other phospholipids.

Immunohistochemistry and immunofluorescence
Mice were deeply anesthetized with pentobarbital (50 mg/kg) and transcardially perfused with phosphate-buffered saline (PBS) followed by 4% paraformaldehyde. The brain was removed, postfixed for 24 h, and transferred successively to 20% and 30% sucrose solutions. Subsequently, brains were frozen on a cold stage and sliced into 30 μm in thickness using cryostat. The sections were permeabilized with 0.3% Triton-X 100 (Sigma-Aldrich, St. Louis, MO) in PBS for 15 min and blocked in PBS containing 1% BSA and 5% normal donkey serum (Jackson ImmunoResearch Lab., West Grove, PA) for 60 min at room temperature. Sections were incubated in the blocking solution (Block Ace, Dainippon Pharmaceutical, Japan) for 30 min at room temperature, and then incubated with rabbit polyclonal antibody against Iba-1 (1:10000; Wako Pure Chem. Indus., Osaka, Japan), which is known to have a specific affinity to the microglial Ca^{2+}-binding protein, and is highly expressed by activated microglia in 10% Block Ace in PBS, at 4 °C overnight. Other sections were incubated with antibody against $Aβ_{1–16}$ (1:1000; Abcam, Cambridge, UK) and NeuN (1:1000; Millipore, Billerica, MA). The rinsed sections were incubated for 6 h with Alexa Fluor 488 goat antirabbit IgG or Alexa Fluor 568 goat antimouse IgG (1:1000; Invitrogen, Eugene, OR) at room temperature. Every treatment was followed by washing three times for 5 min with PBS. Sections were then mounted in the perma fluor aqueous mounting medium (Thermo Fisher Scientific, Waltham, MA).

Quantitative analysis of fluorescence intensity

All samples were analyzed with a confocal laser-scanning microscope (LSM510 Meta; Carl Zeiss, Germany). The number of glial cells in 60–80 areas of 200 μm × 200 μm in five slices per brain was counted and the averaged number/4×10^4 μm² for each brain was obtained.

Measurement of plasmalogen contents in the hippocampus and PFC

Mice were deeply anesthetized with pentobarbital and (50 mg/kg) and transcardially perfused with sterile PBS. The brain was removed and hippocampus was dissected in a dish filled with ice-cold PBS. The samples (300–500 mg) were stored at −80 °C until plasmalogen measurement. The total lipids was extracted by the method of Folch *et al.*,[22] and rel-ative composition of phospholipid classes, including Pls, was measured as reported previously.[21]

Statistical analysis

The results are expressed as the mean ± SEM. The number of Iba-1[+] cells were compared with one-way analysis of variance (ANOVA) followed by a *post hoc* (Scheffe's) test. Changes in PlsEtn contents and the ratio of PlsEtn/Phosphatidyl Etn (PEtn) after LPS and Pls injection were evaluated by the nonpara-metric Kruskal–Wallis test, which was followed by the Steel test for multiple comparisons. Values of $P < 0.05$ were considered statistically significant.

Results

Suppression of glial activation by Pls

As shown in Figure 1A-a, the control group that received saline and corn oil for 7 days showed

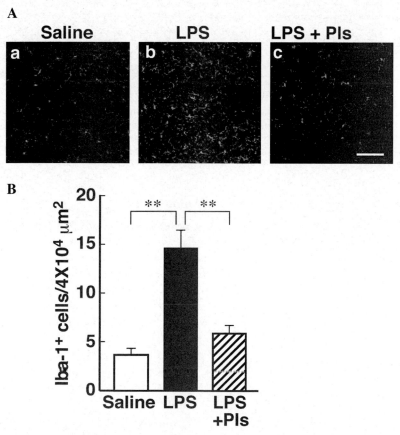

Figure 1. Activation of glial cells following LPS (i.p.) and suppression by simultaneous application of Pls in the mouse hippocam-pus. (A) Number of Iba-1–positive microglia and intensity of immunoreactivity (a, green) increased with LPS treatment (b), which was suppressed by simultaneous application of Pls (c). Scale bar, 100 μm. (B) Summary of the LPS-induced increases in number of microglia, and suppression by Pls (each bar, n = 8). **$P < 0.01$.

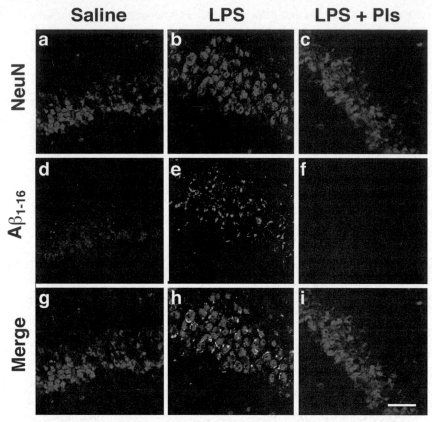

Figure 2. Accumulation of the Aβ protein following LPS (i.p.) and suppression by simultaneous application of Pls in the mouse CA1 region of the hippocampus. Neurons were stained with NeuN (red, a–c). A slight fluorescence for $A\beta_{1-16}$ immunoreactivity (green) of the control group (d) increased following LPS treatment (e), which completely abolished by Pls (f). $A\beta_{1-16}$ fluorescence merged with NeuN immunoreactivity, indicating an intracellular localization of Aβ (g and h). Scale bar, 50 μm.

typical features of Iba-1–positive (green) resting microglia with a small and compact soma bearing ramified processes in the hippocampus. However, itraperitoneal administration of LPS (250 μg/kg/day) for 7 days (LPS group) resulted in neuroinflammation, with an increased number of microglia and intense immunoreactivity and the activated phenotype of cellular hypertrophy and retraction of cytoplasmic processes (Fig. 1A-b). As shown in Figure 1A-a, the increase in activated microglia was suppressed by simultaneous administration of Pls (20 mg/kg) (LPS + Pls group). Figure 1B shows the summary of the LPS-induced increase in the number of glial cells and suppression by Pls (each bar, $n = 8$). One-way ANOVA indicated significant differences in the number of microglia (Fig. 1B; F $(2,21) = 44.7, P < 0.01$) among groups. The *post hoc* test indicated that the LPS group was different from the control and LPS + Pls groups (Scheffe's test,

$P < 0.01$), but the LPS + Pls group was not different from control.

Suppression of LPS-induced Aβ accumulation by Pls

As shown in Figure 2, a slight fluorescence for $A\beta_{1-16}$ immunoreactivity (green) in the CA1 pyramidal cells of the hippocampus in the control group (Fig. 2d) was apparently increased in the LPS group (Fig. 2e), which was completely abolished in the LPS + Pls group (Fig. 2f). Pyramidal neurons were stained with NeuN (red, Figs. 2a–c), and most of $A\beta_{1-16}$ fluorescence merged with NeuN immunoreactivity, indicating an intracellular localization of Aβ (yellow in Fig. 2h).

Changes in plasmalogen contents in the brain after LPS and Pls treatment

As shown in Table 1, PlsEtn was much higher than PlsCho levels in the hippocampus. Relative content

Table 1. Changes in phospholipid content in the hippocampus after LPS and Pls injection

	Control	LPS	LPS + Pls
PlsEtn	23.98 ± 0.80	20.49 ± 0.44*	22.28 ± 0.70
PEtn	19.98 ± 0.47	21.87 ± 0.76	20.41 ± 0.51
PlsCho	0.51 ± 0.09	0.40 ± 0.03	0.47 ± 0.05
PCho	36.70 ± 1.63	39.19 ± 0.53	37.84 ± 1.43
SM	5.44 ± 0.64	5.16 ± 0.44	5.51 ± 0.68
PS	11.93 ± 0.79	11.14 ± 0.44	11.91 ± 0.710
PI	1.65 ± 0.12	1.83 ± 0.08	1.74 ± 0.07

PlsEtn, ethanolamine plasmalogen; PEtn, phosphatidyl ethanolamine; PlsCho, choline plasmalogen; PCho, phosphatidyl choline; SM, sphingomyelin; PS, phosphatidyl serine; PI, phosphatidyl inositol. Values are the means ± SEM ($n = 5$) of % contents of phospholipids, except for the last row. *$P < 0.05$ compared with the control group (Steel test for multiple comparisons).

of PlsEtn was significantly decreased by LPS injection (Kruskal–Wallis test, $\chi^2(2) = 7.74$, $P < 0.05$). Multiple comparison analysis by the Steel test revealed that PlsEtn content in the LPS group, but not in LPS + Pls, significantly decreased compared with the control group ($P < 0.05$, each group, $n = 5$). Contents of other phospholipids showed no significant changes following any treatments.

Discussion

One of the possible mechanisms of the LPS-induced amyloidogenesis that has been recently suggested is that the activity of β- and γ-secretase, which are deeply involved in the amyloidogenic processing from amyloid precursor protein,[23] increases in the cortex and hippocampus following systemic injection of LPS.[2] It has been shown that proinflammatory cytokines, as well as ROS/RNS, released from activated microglia augment Aβ formation through upregulating β-secretase mRNA and enzymatic activity.[5,24] Microglia are activated further through receptors for advanced glycation end product (RAGE), which bind Aβ, or induce phagocytosis of Aβ, thereby amplifying generation of ROS/RNS and cytokines.[23]

In addition to the activation of glial cells and Aβ accumulation, here we showed for the first time that plasmalogen contents in the hippocampus deceased following LPS administration (Table 1). A possible mechanism of the decrease in Pls during neuroinflammation may result from an antioxidant property of Pls that protects cells from oxidative stress. It has been shown that Pls-specific vinyl ether bond at the *sn*-1 of the glycerol backbone is targeted by a vast variety of oxidants, including ROS/RNS,[10,11,13] and oxidative stress preferentially oxidizes PlsEtn over phosphatidyl ethanolamine,[25,26] resulting in the disruption of vesicular fusion in the synaptosomes and acetylcholine release.[27] This may at least partly explain why AD patients show a decrease in Pls content in the brain.[14–16] It has been suggested that an abnormal membrane lipid composition, namely, a decrease in the ratio of Pls to non-Pls ethanolamine glycerephospholipids, causes membrane instability in AD, which may contribute to amyloidogensis by cooperatively acting with amyloid cascade mechanism.[14] Furthermore, since PlsEtn are major endogenous lipid constituents that facilitate membrane fusion of synaptic vesicles associated with neurotransmitter release,[28,29] age-related and/or pathological alteration in Pls content is suggested to play important roles in neurological disorders including AD.[7] In fact, the decrease in Pls is shown to be closely correlated with the severity of dementia in humans.[18,19]

Several lines of evidence have suggested that there is a causal loop among neuroinflammation, Aβ accumulation, ROS/RNS production, and decrease in Pls. The LPS-induced activation of β-secretase,[2] which is predominantly localized in cholesterol-rich lipid raft,[30,31] causes accumulation of the Aβ protein. The Aβ-induced production of ROS/RNS enhances lipid peroxidation,[32,33] resulting in the decrease in Pls content, as mentioned previously. In addition, it has been shown that increased Aβ and ROS/RNS reduce expression of alkyl-dihydroxyacetoneposphate-synthase, a rate-limiting enzyme for plasmalogen *de novo* synthesis, due to the dysfunction of peroxisomes where Pls are biosynthesized, resulting in the decrease in plasmalogen level.[34] It is also reported that the LPS-induced neuroinflammation itself can cause the loss of Pls by inducing TNF-α that downregulates another key enzyme in plasmalogen biosynthesis in peroxisomes, glycerol-3-phosphate-O-acyltransferase,[35] and by upregulating myeloperoxidase that generates one of the reactive species, hypochlorous acid (HOCl), in the brain, targeting Pls to be oxidized.[36] Finally, Pls-PLA$_2$, which degrades Pls to release DHA or ARA from the *sn*-2 position of the glycerol moiety, is possibly activated by ceramide produced under inflammatory conditions and contributes to the loss of Pls in the brain.[9,37]

The reduction of Pls has been shown to induce a decreased ability to intracellular cholesterol transport from cell membrane to the endoplasmic reticulum, resulting in an increase in cholesterol on the cell surface.[38] On the other hand, it is well known that the generation and clearance of Aβ are affected by cholesterol metabolism, which is evidenced by the identification of a variant gene of the apolipoprotein E, a cholesterol transporter, as a major genetic risk factor for AD.[23,39,40] Depletion of cholesterol inhibits the generation of Aβ,[41,42] while an increase in cholesterol promotes secretion of Aβ.[39,40,43] Thus, a vicious circle of LPS-induced Aβ accumulation that decreases the Pls level, which leads to an increase in cholesterol, and thereby enhances further generation of Aβ, is established under the pathological condition of neuroinflammation.

In this study, we showed that the LPS-induced activation of glial cells (Fig. 1), accumulation of the Aβ protein (Fig. 2), and decrease in PlsEtn (Table 1) in the hippocampus were suppressed by peripheral injection of Pls. The precise mechanism of Pl effects is not known in this study. Since glial activation, Aβ accumulation, and Pls reduction are all involved in the vicious circle mentioned previously, it is possible that effective supplementation of Pls can improve all of these pathological disorders. The most important question may be whether peripheral Pls can enter the brain. There are no reports thus far that Pls directly cross the blood–brain barrier (BBB). Therefore, it cannot be excluded that the antioxidative effects of Pls are exerted outside the brain to suppress the primary inflammation induced by peripheral LPS. However, it has been shown that Pls in the serum decrease in parallel with, or even earlier than, the decrease in brain Pls in AD patients.[18,19] Furthermore, our results showed that the LPS-induced decrease in plasmalogen contents in the hippocampus was corrected by the peripheral administration of Pls (Table 1). Therefore, it is possible that a peripheral supplementation of Pls could be expected to have an effect on the central nervous system (CNS).

Another question is whether the effective molecules in our experiment are not Pls themselves, but polyunsaturated fatty acids (PUFAs), which Pls must carry at the *sn*-2 position. There are several lines of evidence showing that *n*-3 PUFAs, such as eicosapentaenoic acid, DHA, and its derivative, neuroprotectin D1, have anti-inflammatory and neuroprotective effects.[44–47] Furthermore, DHA has been reported to suppress production of the Aβ protein through multiple mechanisms including inhibition of β-/γ-secretase activities and alteration of membrane cholesterol distribution.[48–50] Since the purified Pls used in this study contain DHA and its precursor, α-linolenic acid, especially in PlsEtn, it is possible that DHA derived from PlsEtn plays a significant role in the CNS effects of Pls. In fact, it has been shown that DHA is synthesized from α-linolenic acid and incorporated into phospholipids in the liver, and is then transported to the brain through the peripheral circulation.[51] On the other hand, it has been shown that lyso-type phospholipids, which contain DHA at the *sn*-2, show preferential transfer through an *in vitro* model of the BBB over DHA.[52] This finding suggests that Pls-containing DHA are more effective to exert the CNS actions than DHA itself.

This study suggests that peripheral administration of Pls, of which extraction techniques have been recently developed,[20,21] may suppress neuroinflammation in the brain. Although further investigations concerning mechanisms of the CNS effects are needed, including the pathways for entering into the brain, the present results indicate the possibility of Pls as a new preventive/therapeutic strategy for Alzheimer's disease.

Acknowledgment

This work was supported by Grants-in-Aid for Scientific Research (22590225) to T.K. from the Japanese Ministry Education, Culture, Sports, Science, and Technology.

Conflicts of interest

The authors declare no conflicts of interest.

References

1. Boehm, G.W. *et al.* 2005. Effects of intraperitoneal lipopolysaccharide on Morris maze performance in year-old and 2-month-old female C57BL/6J mice. *Behav. Brain Res.* **159:** 145–151.

2. Lee, J.W., *et al.* 2008. Neuro-inflammation induced by lipopolysaccharide causes cognitive impairment through enhancement of β-amyloid generation. *J. Neuroinflammation* **5:** 37.

3. O'mara, S.M., K.N. Shaw & S. Commins. 2001. Lipopolysaccharide causes deficits in spatial learning in the watermaze but not in BDNF expression in the rat dentate gyrus. *Behav. Brain Res.* **124:** 47–54.

4. Blasko, I., *et al.* 2001. Ibuprofen decreases cytokine-induced amyloid β production in neuronal cells. *Neurobiol. Dis.* **8:** 1094–1101.

5. Sastre, M., *et al.* 2003. Nonsteroidal anti-inflammatory drugs and peroxisome proliferator-activated receptor-γ agonists modulate immunostimulated processing of amyloid precursor protein through regulation of β-secretase. *J. Neurosci.* **23:** 9796–9804.

6. Yan, Q., *et al.* 2003. Anti-inflammatory drug therapy alters β-amyloid processing and deposition in an animal model of Alzheimer's disease. *J. Neurosci.* **23:** 7504–7509.

7. Farooqui, A.A. & L.A. Horrocks. 2001. Plasmalogens: workhorse lipids of membranes in normal and injured neurons and glia. *Neuroscientist* **7:** 232–245.

8. Farooqui, A.A. & L.A. Horrocks. 2001. Plasmalogens, phospholipase A2, and docosahexaenoic acid turnover in brain tissue. *J. Mol. Neurosci.* **16:** 263–272; discussion 279–284.

9. Farooqui, A.A. 2010. Studies on plasmalogen-selective phospholipase A2 in brain. *Mol. Neurobiol.* **41:** 267–273.

10. Khaselev, N. & R.C. Murphy. 1999. Susceptibility of plasmenyl glycerophosphoethanolamine lipids containing arachidonate to oxidative degradation. *Free Radic. Biol. Med.* **26:** 275–284.

11. Engelmann, B. 2004. Plasmalogens: targets for oxidants and major lipophilic antioxidants. *Biochem. Soc. Trans.* **32:** 147–150.

12. Maeba, R. & N. Ueta. 2003. Ethanolamine plasmalogen and cholesterol reduce the total membrane oxidizability measured by the oxygen uptake method. *Biochem. Biophys. Res. Commun.* **302:** 265–270.

13. Yavin, E. & S. Gatt. 1972. Oxygen-dependent cleavage of the vinyl -ether linkage of plasmalogens: 2. Identification of the low-molecular-weight active component and the reaction mechanism. *Eur. J. Biochem.* **25:** 437–446.

14. Ginsberg, L., *et al.* 1995. Disease and anatomic specificity of ethanolamine plasmalogen deficiency in Alzheimer's disease brain. *Brain Res.* **698:** 223–226.

15. Guan, Z., *et al.* 1999. Decrease and structural modifications of phosphatidylethanolamine plasmalogen in the brain with Alzheimer disease. *J. Neuropathol. Exp. Neurol.* **58:** 740–747.

16. Han, X., D.M. Holtzman & D.W. McKeel, Jr. 2001. Plasmalogen deficiency in early Alzheimer's disease subjects and in animal models: molecular characterization using electrospray ionization mass spectrometry. *J. Neurochem.* **77:** 1168–1180.

17. Ginsberg, L., J.H. Xuereb & N.L. Gershfeld. 1998. Membrane instability, plasmalogen content, and Alzheimer's disease. *J. Neurochem.* **70:** 2533–2538.

18. Goodenowe, D.B., *et al.* 2007. Peripheral ethanolamine plasmalogen deficiency: a logical causative factor in Alzheimer's disease and dementia. *J. Lipid Res.* **48:** 2485–2498.

19. Wood, P.L., *et al.* 2010. Circulating plasmalogen levels and Alzheimer disease assessment scale-cognitive scores in Alzheimer patients. *J. Psychiat. Neurosci.* **35:** 59–62.

20. Mawatari, S., *et al.* 2009. Simultaneous preparation of purified plasmalogens and sphingomyelin in human erythrocytes with phospholipase A1 from Aspergillus orizae. *Biosci. Biotechnol. Biochem.* **73:** 2621–2625.

21. Mawatari, S., Y. Okuma & T. Fujino. 2007. Separation of intact plasmalogens and all other phospholipids by a single run of high-performance liquid chromatography. *Anal. Biochem.* **370:** 54–59.

22. Folch, J., M. Lees & G.H. Sloane Stanley. 1957. A simple method for the isolation and purification of total lipides from animal tissues. *J. Biol. Chem.* **226:** 497–509.

23. Querfurth, H.W. & F.M. LaFerla. 2010. Alzheimer's disease. *N. Engl. J. Med.* **362:** 329–344.

24. Blasko, I., *et al.* 1999. TNFα plus IFNγ induce the production of Alzheimer β-amyloid peptides and decrease the secretion of APPs. *FASEB J.* **13:** 63–68.

25. Reiss, D., K. Beyer & B. Engelmann. 1997. Delayed oxidative degradation of polyunsaturated diacyl phospholipids in the presence of plasmalogen phospholipids in vitro. *Biochem. J.* **323**(Pt 3): 807–814.

26. Zoeller, R.A., *et al.* 2002. Increasing plasmalogen levels protects human endothelial cells during hypoxia. *Am. J. Physiol. Heart Circ. Physiol.* **283:** H671–H679.

27. Urano, S., *et al.* 1997. Oxidative injury of synapse and alteration of antioxidative defense systems in rats, and its prevention by vitamin E. *Eur. J Biochem./FEBS* **245:** 64–70.

28. Lohner, K., *et al.* 1991. Stabilization of non-bilayer structures by the etherlipid ethanolamine plasmalogen. *Biochim. Biophys. Acta* **1061:** 132–140.

29. Breckenridge, W.C., *et al.* 1973. Adult rat brain synaptic vesicles. II. Lipid composition. *Biochim. Biophys. Acta* **320:** 681–686.

30. Cordy, J.M., *et al.* 2003. Exclusively targeting β-secretase to lipid rafts by GPI-anchor addition up-regulates β-site processing of the amyloid precursor protein. *Proc. Natl. Acad. Sci. U.S.A.* **100:** 11735–11740.

31. Riddell, D.R., *et al.* 2001. Compartmentalization of β-secretase (Asp2) into low-buoyant density, noncaveolar lipid rafts. *Curr. Biol.* **11:** 1288–1293.

32. Butterfield, D.A. & C.M. Lauderback. 2002. Lipid peroxidation and protein oxidation in Alzheimer's disease brain: potential causes and consequences involving amyloid β-peptide-associated free radical oxidative stress. *Free Radic. Biol. Med.* **32:** 1050–1060.

33. Christen, Y. 2000. Oxidative stress and Alzheimer disease. *Am. J. Clin. Nutr.* **71:** 621S–629S.

34. Grimm, M.O.W., *et al.* 2011. Plasmalogen synthesis is regulated via alkyl-dihydroxyacetonephosphate-synthase by amyloid precursor protein processing and is affected in Alzheimer's disease. *J. Neurochem.* **116:** 916–925.

35. Cimini, A., *et al.* 2003. TNFα downregulates PPARδ expression in oligodendrocyte progenitor cells: implications for demyelinating diseases. *Glia* **41:** 3–14.

36. Üllen, A., *et al.* 2010. Mouse brain plasmalogens are targets for hypochlorous acid-mediated modification in vitro and in vivo. *Free Rad. Biol. Med.* **49:** 1655–1665.

37. Latorre, E., *et al.* 2003. Signaling events mediating activation of brain ethanolamine plasmalogen hydrolysis by ceramide. *Eur. J. Biochem.* **270:** 36–46.

38. Munn, N.J., *et al.* 2003. Deficiency in ethanolamine plasmalogen leads to altered cholesterol transport. *J. Lipid Res.* **44:** 182–192.

39. Frears, E.R., *et al.* 1999. The role of cholesterol in the biosynthesis of β-amyloid. *Neuroreport* **10:** 1699–1705.

40. Puglielli, L., R.E. Tanzi & D.M. Kovacs. 2003. Alzheimer's disease: the cholesterol connection. *Nat. Neurosci.* **6:** 345–351.

41. Simons, M., *et al.* 1998. Cholesterol depletion inhibits the generation of β-amyloid in hippocampal neurons. *Proc. Natl. Acad. Sci. U.S.A.* **95:** 6460–6464.

42. Kojro, E., *et al.* 2001. Low cholesterol stimulates the nonamyloidogenic pathway by its effect on the α-secretase ADAM 10. *Proc. Natl. Acad. Sci. U.S.A.* **98:** 5815–5820.

43. Chauhan, N.B. 2003. Membrane dynamics, cholesterol homeostasis, and Alzheimer's disease. *J. Lipid. Res.* **44:** 2019–2029.

44. Farooqui, A.A., L.A. Horrocks & T. Farooqui. 2007. Modulation of inflammation in brain: a matter of fat. *J. Neurochem.* **101:** 577–599.

45. Calon, F., *et al.* 2004. Docosahexaenoic acid protects from dendritic pathology in an Alzheimer's disease mouse model. *Neuron* **43:** 633–645.

46. Palacios-Pelaez, R., W.J. Lukiw & N.G. Bazan. 2010. Omega-3 essential fatty acids modulate initiation and progression of neurodegenerative disease. *Mol. Neurobiol.* **41:** 367–374.

47. Schmitz, G. & J. Ecker. 2008. The opposing effects of *n*-3 and *n*-6 fatty acids. *Prog. Lipid Res.* **47:** 147–155.

48. Grimm, M.O.W., *et al.* 2011. Docosahexaenoic acid reduces amyloid beta production via multiple pleiotropic mechanisms. *J. Biol. Chem.* **286:** 14028–14039.

49. Walsh, D.M. & D.J. Selkoe. 2004. Deciphering the molecular basis of memory failure in Alzheimer's disease. *Neuron* **44:** 181–193.

50. Oksman, M., *et al.* 2006. Impact of different saturated fatty acid, polyunsaturated fatty acid and cholesterol containing diets on β-amyloid accumulation in APP/PS1 transgenic mice. *Neurobiol. Dis.* **23:** 563–572.

51. Scott, B.L. & N.G. Bazan. 1989. Membrane docosahexaenoate is supplied to the developing brain and retina by the liver. *Proc. Natl. Acad. Sci. U.S.A.* **86:** 2903–2907.

52. Bernoud, N., *et al.* 1999. Preferential transfer of 2-docosahexaenoyl-1-lysophosphatidylcholine through an in vitro blood-brain barrier over unesterified docosahexaenoic acid. *J. Neurochem.* **72:** 338–345.

Ann. N.Y. Acad. Sci. ISSN 0077-8923

ANNALS OF THE NEW YORK ACADEMY OF SCIENCES

Issue: *Neuroimmunomodulation in Health and Disease*

Biological memory of childhood maltreatment: current knowledge and recommendations for future research

Katharina Schury and Iris-Tatjana Kolassa

Institute of Psychology and Education, Clinical and Biological Psychology, University of Ulm, Ulm, Germany

Address for correspondence: Katharina Schury, University of Ulm, Institute of Psychology and Education, Clinical and Biological Psychology, Albert-Einstein-Allee 47, 89069 Ulm, Germany. katharina.schury@uni-ulm.de

Child maltreatment (CM) not only has detrimental and lifelong psychological consequences, but also can lead to lasting alterations in core physiological systems—a biological memory of CM. Furthermore, some of these alterations might even be transmitted to the next generation. This article describes current knowledge about the effects of CM on the stress system (i.e., the hypothalamus–pituitary–adrenal axis), on cellular aging (i.e., telomere length and telomerase activity), and on the immune system. Furthermore, we want to initiate research on the question of transmission of the described physiological alterations subsequent to CM to the next generation—possibly through epigenetic imprinting. As diverse neurobiological factors and epigenetics are closely linked, these different research fields should join forces to gain a deeper understanding of the biological determinants and sequelae of CM and its transmission.

Keywords: childhood maltreatment; epigenetic; telomere length; immune system; HPA axis

Introduction

Child maltreatment (CM) has detrimental effects on children's development with psychobiosocial consequences throughout life. Furthermore, CM—and its consequences—can be transmitted to the next generation, leading to a transgenerational cycle of maltreatment (TCM). This article aims at describing existing knowledge on neurobiological determinants and sequelae of CM—a biological memory of CM. Furthermore, we will respond to possible transgenerational effects of CM. We do not attempt to systematically review all of the work in this widespread field. Rather, we want to connect knowledge from different fields of research—stress system, immune system, cell aging, and epigenetics—to stimulate research on this innovative and societally important topic.

CM includes sexual, physical, and emotional abuse, as well as physical and emotional neglect (cf. Refs. 1–3), and there is strong evidence that different forms of maltreatment are interrelated.[4,5] Furthermore, the number of different types of maltreatment (e.g., beating, verbal humiliation) might

be critical in the development of consequential dysfunctions. We know from posttraumatic stress disorder (PTSD)—a psychiatric condition that can develop after traumatic experiences—that the risk for developing the disorder increases with the number of different traumatic event types experienced, that is, with increasing traumatic load.[6–8] Similarly, in CM, a cumulative effect on developmental problems during adolescence and health in adult life can be supposed[9,10]—an effect that could be described as *maltreatment load*.

Prevalence rates for maltreatment vary considerably due to methodological issues, like differing definitions of maltreatment, variations in assessment methods (e.g., agency report, self, or parent report), or samples used, but also due to high estimated numbers of unreported cases. Cumulative prevalence rates between 5–35% are reported for physical abuse, 4–9% for severe emotional abuse, 15–30% for sexual abuse in girls and 5–15% in boys, and 6–11.8% for neglect (cf., Ref. 11).

Common consequences of CM are mental health problems, such as PTSD, depression, and anxiety, as well as cognitive distortions, risky sexual behavior,

doi: 10.1111/j.1749-6632.2012.06617.x

obesity, and criminal behavior.[11,12] Furthermore, maltreated children are reported to have poorer physical health throughout life, including cardiovascular disease, diabetes, cancer, frequent headaches, accelerated cell aging, and even premature mortality.[13–15] Maltreatment may be especially detrimental in early childhood, as this time period is characterized by rapid neuroendocrinological development, which renders the brain and stress axes especially vulnerable to the influence of stress.[16]

Children of parents with a history of CM are at increased risk to experience CM themselves,[17–20] leading to a transgenerational cycle of maltreatment, with transmission rates around 7%.[17,21] However, importantly, the majority of parents with their personal CM experiences do not maintain the cycle of maltreatment.[17,22] Several risk and protective factors are known to play a role in TCM. Poor parenting styles, combined with parenting at under 21 years, history of mental illness, and residing with a violent adult put families at increased risk for TCM,[23] whereas financial stability, social support, a caring and stable caregiver appear to be protective.[17,24]

Identifying pre- and perinatal factors that play a role in the determinants and sequelae of CM, as well as in the transmission of maltreatment to the next generation, is an important step that will finally help to advance interventions and services to families and children at risk.

Biological mechanisms in CM-related disorders and in TCM

To understand the development of a biological memory of CM as well as the transmission of CM, mediating biological factors have to be taken into account. In this paper, the state of research on alterations in the neurobiological systems (i.e., HPA axis, biomarkers of cell aging, and the immune system), which are involved in the etiology of CM-related dysfunctions and in the transmission of maltreatment as well as epigenetic mechanisms in TCM, will be characterized. A brief introduction to these biological systems will be given, followed by a narrative review of studies concerning the association of the particular system and CM as well as TCM, respectively.

HPA axis
The HPA axis is a key factor in the response to stress and leads to the release of the stress hormone cor-

tisol. In response to stress, corticotrophin-releasing hormone (CRH) is released from the hypothalamus, which stimulates the release of adrenocorticotropin releasing hormone (ACTH) in the pituitary gland, resulting in the release of glucocorticoids (mostly cortisol) from the adrenal gland into blood circulation. The activity of the HPA axis is subject to a predictable circadian cycle and regulates itself by negative feedback loops, that is, CRH and ACTH release is inhibited when levels of blood cortisol are high. Temporarily elevated cortisol levels in response to an acute stressor are important for the organisms' adaption to environmental stressors and are needed to maintain homeostasis. However, childhood adversity seems to alter HPA axis activity, and these early changes persist.[25] Chronically altered levels of cortisol increase the risk for later psychopathology like depression (which is associated with hypercortisolism[26]) or PTSD (which is associated with hypocortisolism in patients with PTSD due to physical or sexual abuse[27]), but also for stress-related diseases like asthma or rheumatoid arthritis.[28] In this context, the suppressing effect of cortisol on the immune system has to be mentioned, which can result in excessive inflammation.

Early life stress has been linked to abnormal functioning of the HPA axis in several investigations. But evidence of alterations in the HPA axis in the aftermath of CM is inconsistent (for a review, see Ref. 25). The conflicting results on HPA axis alterations after CM might stem from methodological issues like the operationalization of maltreatment, the sample (e.g., psychiatric patients vs. general population), or the collection and measurement of cortisol.[29] So far, cortisol has mostly been measured in urine, blood, or saliva. These measurements allow the assessment of cortisol over short time periods (minutes in blood or saliva, hours in saliva), but are subject to situational factors (e.g., food intake, perceived stress). In particular, the circadian rhythm of cortisol production makes it difficult to compare results, and the samples should be taken in accordance with a strict protocol. A new and promising approach to overcome these methodological problems is the analysis of cortisol in hair, in which cortisol levels can be estimated over longer time periods (approximately three months).[30] The differing results on HPA axis activity subsequent to CM and/or trauma might also mirror the influence of other hormones, which are known to inhibit cortisol. The hormone dehydroepiandrosterone (DHEA) is a cortisol antagonist and protects the brain—in particular the hippocampus—from the damaging effect of cortisol and inhibits excessive inflammatory processes resulting from cortisol.[31] When estimating HPA axis activity, one should not only rely on cortisol levels, but also should consider using the ratio of cortisol and the cortisol antagonist DHEA. Studies have shown that the cortisol/DHEA ratio can serve as a more reliable marker of HPA axis activity (for an overview, see Ref. 31).

Despite conflicting results, one can reason that CM leads to a dysregulation of the HPA axis, thereby impairing stress reactivity as well as health (through the influence of cortisol on inflammation) throughout the life of affected individuals. Even worse, alterations of the HPA axis might be transmitted to the next generation leading to a potential vulnerability of the infant's stress system, but there are only a few studies that address transgenerational effects. A 25-year longitudinal study observed lower morning cortisol as well as altered cortisol release during the course of the day in control children compared with children of mothers that reported social withdrawal in their childhood.[32] As no cortisol levels of the mothers were assessed, developmental similarities in stress response patterns between mothers and their children can be supposed to exist but not confirmed in this study. In another study, lower baseline cortisol in infants of CM mothers were observed compared with control mother–infant pairs.[33] Transgenerational effects of maternal stress caused by mental disorder on the HPA axis of their infants have also been confirmed in mothers with depression[34,35] or PTSD.[36,37]

The HPA axis is the core system for stress regulation and affects other important physiological systems, for example, the immune system. An effect of maternal CM on the infants, HPA axis would increase the risk for impaired stress adaption and stress-related illnesses in these children from the very beginning of their life. We need to understand the mechanisms of the transmission of altered HPA axis activity and develop interventions to circumvent the way in which experiences of maltreatment not only affect the life of a mother negatively, but also affect her children. More studies on the transmission of HPA axis dysregulations to the next generation are needed. These studies should try to overcome methodical problems, for example, by strict measurement protocols, using the cortisol/DHEA ratio to estimate HPA axis activity, or by considering new methods like the analysis of cortisol in hair.

Cell aging

Chronic and severe psychological stress can lead to premature cell aging, measured by telomere length and telomerase activity.[38] Telomeric DNA is composed of repetitive DNA sequences (in humans TTAGGG); it caps the end of chromosomes and promotes chromosomal stability. In the process of somatic cell division, the telomere is not fully replicated and thus shortened at each replication,[39] finally leading to programmed cell death or senescence.[40] The enzyme telomerase adds the necessary telomeric DNA and has direct telomere-protective functions.[41] Telomere length and telomerase activity are accepted biomarkers for cell aging and longevity, and shortened telomeres have been linked to several age-related disorders like cardiovascular disease, cancer, diabetes,[42,43] and even all-cause mortality.[44]

In one of the first investigations on the effects of psychological stress on cell aging, women with high rates of chronic stress showed lower telomerase activity and accelerated telomere shortening compared with women with low stress levels.[38] Shortened telomeres have in the meantime been confirmed in the aftermath of CM as well: one study observed shorter telomere length in mentally healthy adults with a history of maltreatment compared with nonmaltreated controls.[45] In particular, neglect or other adverse childhood experiences (ACE), such as parents' divorce or death of parent, which are usually seen as less detrimental than physical or sexual abuse, can affect health through telomere shortening in adult life, with a difference of 640 bp compared with individuals without adverse childhood experiences.[46] With an average shortening rate of 31–63 bp per year (average shortening rate across adulthood from 20–95 years; cf. Ref. 38), this can translate into a 7- to 15-year difference in life span. Adults (with and without an anxiety disorder) who reported ACE showed shorter telomeres than adults without a history of ACE.[47] Furthermore, a recent study on socially deprived children confirmed that accelerated telomere shortening can already be detected during childhood.[48] O'Donovan et al.[49] observed reduced telomere length subsequent to CM in PTSD patients. Importantly, the number of different categories of traumatic events in childhood was linearly associated with accelerated telomere shortening, indicating a cumulative effect of childhood trauma (maltreatment load). CM also accounted for the effect of PTSD on telomere length in this study. Only one study so far failed to confirm an association between CM and shorter telomere length.[50] This might be due to the rather superficial assessment of CM, as participants were asked only five questions regarding whether they had experienced physical, emotional, or sexual abuse in childhood or a later time period. In addition, the study included only participants who were twins.

Research on accelerated telomere shortening in CM is still sparse, and more research on the underpinnings of telomere shortening and possible transgenerational effects is needed. The role of telomerase activity has not yet been investigated in CM.

Immune system

The immune system protects the body from pathogenic organisms like bacteria, viruses, fungus, or defective cells of the own body. After tissue injury, but also in response to stress, inflammatory processes are implemented by the immune system via the release of pro- and anti-inflammatory cytokines. Inflammatory processes, however, have also been linked to the development of several chronic diseases such as cardiovascular, autoimmune, mental, or neurodegenerative diseases.[51–53]

A large longitudinal study confirmed long-term effects of CM on adult inflammation. Individuals with a history of childhood maltreatment or social isolation had an increased risk for heightened inflammation (C-reactive protein level > 3 mg/L) in adulthood, independent of the influence of additional early life stress, stress in adulthood or adult health, and health behavior in this prospective study.[54,55] In a study with 132 healthy older adults, including 58 dementia family caregivers, a history of multiple childhood adversities or childhood abuse was associated with elevated levels of proinflammatory cytokines.[46] The reported relationship was more pronounced in caregivers compared with noncaregivers. Another study showed a greater overall peripheral release of the proinflammatory cytokine interleukin (IL)-6 and elevated IL-6 concentrations during a stress test in individuals with CM, compared with nonmaltreated controls.[56]

Potentially, traumatic experiences might also alter immune function via epigenetic changes. For example, Uddin et al.[77] demonstrated that a greater number of unmethylated genes led to changes in gene expression, which were related to altered immune function.

We hypothesize that CM can also influence the immune system of the next generation—possibly via epigenetic pathways—leading to higher risks for illnesses in children of mothers with CM. One study showed that children of mothers who were exposed to chronic interpersonal violence (during childhood as well as later on) had elevated IgE levels,[57] which are associated with the development of allergies in later life.[58] The mechanism underlying this transgenerational effect of CM on the immune system was not investigated in this study. We assume that epigenetic pathways can explain such phenomena. Immunological "knowledge" seems to be epigenetically passed from the mother to her infant during limited neonatal imprinting periods (discussed further below), as well as during gestation, with generally beneficial effects for the child's immune system.[59–61] Despite the theoretical knowledge about such epigenetic transmission processes, very few studies address transgenerational pathways of immune alterations in CM.

The role of epigenetic alterations in maltreatment, trauma, and the transmission of both

Usually when speaking of "inheritance" one refers to the passing of genetic information from one generation to the next.[62] However, not only the presence of genes but the level of gene expression determine individual variations in offspring phenotype. The level of gene expression can be regulated by genetic polymorphisms, but there is mounting evidence that environmental effects can be transmitted to following generations via epigenetic alterations,[62] as was, for example, described for the immune system in the aforementioned study by Sternthal et al.[57]

Epigenetic mechanisms, that is, alterations in chromatin structure such as DNA methylation, histone modifications, and RNA interference, can occur in response to CM and might be involved in the transgenerational transmission of childhood maltreatment and traumatization.[63,64] There are different pathways by which epigenetic alterations in the offspring of CM mothers can be explained. First, stress during pregnancy might directly alter the epigenome of the child (e.g., Ref. 65). Second, postnatal mother–infant interactions and low maternal care, which can be altered as a consequence of maltreatment experiences of the mother, can change the infant's epigenome. Alterations in the methylation status of specific gene sites were observed in rat pups as a result of low maternal care behavior, compared with high maternal care. This was shown for the glucocorticoid receptor gene promoter in the hippocampus,[66] the glutamic acid decarboxylase 1 (GAD1) promoter in the hippocampus,[67] and the brain-derived neurotrophic factor (BDNF) gene.[68] Cross-fostering seems to reverse the methylation changes associated with less attentive maternal care, emphasizing the ongoing importance of environmental influences and the possibility for interventions.[66] There is also one study that did not find alterations in the methylation status (in the glucocorticoid receptor promoter region) after early life adversity in rat pups, although abnormal behaviors, increased neurotrophin levels, and HPA axis dysregulation were observed.[69] Third, gametic transmission of epigenetic alterations can occur; for example, unpredictable and chronic separation from the mother altered DNA methylation in the promoter region of several candidate genes in the germline of male mice.[70] The epigenome is largely erased during early preimplantation development of the germline and during germ cell development. However, some epigenetic marks are maintained, thus permitting gametic transmission of epigenetic alterations, making transgenerational epigenetic inheritance possible.[71]

Very few studies have investigated transgenerational transmission of epigenetic changes subsequent to traumatic stress in humans. The first human study on this topic investigated influences of stress during pregnancy on the infant's epigenome. Exposure to depressed mood of the mothers in the third trimester of pregnancy was related to increased neonatal methylation levels at the glucocorticoid receptor promoter region.[72] Interestingly, this increase of methylation levels was associated with an elevated stress responses of the HPA axis at three months after birth. Prenatal exposure to maternal depressed mood also affected the methylation patterns at the serotonin-transporter-linked polymorphic region (5-HTTLPR).[73] And maternal gestational exposure to interpartnership violence influenced the methylation status in the promoter of the glucocorticoid receptor in the offspring even 10–19 years after birth.[65] The few studies in humans point out that maternal stress during pregnancy can alter methylation patterns in their infants in affective- and stress-related gene sites (5-HTTLPR, glucocorticoid receptor gene), presumably leading to alterations in neurobiological systems in the infant. Evidence, however, is still sparse and more and larger studies need to be implemented.

In investigating epigenetics, one has to keep in mind that results of epigenetic alterations measured in blood have to be interpreted with caution. Epigenetic changes are cell specific, and so far it is unknown how findings from the blood reflect changes in the brain. One study showed that childhood abuse leads to epigenetic changes in the brain in humans.[74] In this study, alterations in the gene expression in the glucocorticoid receptor were found in the hippocampus of suicide victims with a history of child abuse compared with suicide victims without a history of child abuse and controls. Nevertheless, the issue of relating epigenetic findings in the blood with alterations in the brain remains open and has to be further addressed in humans and in animal models to gain a better understanding of the exact mechanisms of epigenetic transmission.

Epigenetics might be the key for understanding transgenerational transmission processes in CM and trauma. But knowledge from human studies is very limited and there are still methodological limitations that do not allow the generalization of study results (e.g., other influencing factors besides CM cannot be controlled in such studies). Nevertheless, the investigation of epigenetic alterations subsequent to CM and—most importantly—the transmission to the next generation seems to be a worthwhile undertaking.

Connecting neurobiology and epigenetics

The aforementioned neurobiological and epigenetic systems are closely linked to each other and to environmental factors (see Fig. 1). Early life experiences can modulate the stress axis (i.e., the HPA axis), thereby altering basal cortisol levels as well as the release of cortisol after a stressor. The stress response might also be influenced by epigenetic processes, for example, by influencing the glucocorticoid receptor promoter (cf. Ref. 65). Excessive

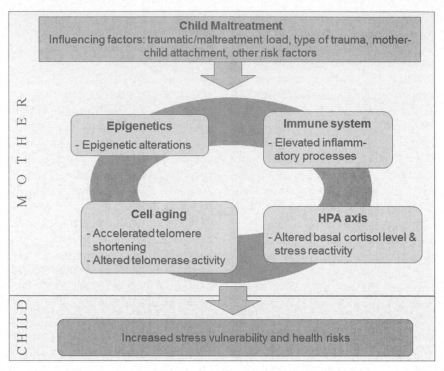

Figure 1. Schematic illustration of the association of core physiological systems that are altered subsequent to child maltreatment (cf. upper panel, mother), as well as the transmission of these alterations to the next generation (cf. lower panel, child). GR = glucocorticoid receptor.

activation of the body's stress systems can also accelerate telomere shortening, and thus, premature cell aging. Furthermore, the stress hormone cortisol is immunosuppressive. However, under severe stress, cortisol can potentiate, instead of inhibit, the activity of IL-1 and IL-6,[75] leading to elevated inflammation by impairing the functioning and regulation of the immune response to stress. Stress-related activation of the immune system might also be linked to telomere shortening and thereby an earlier onset of age-related diseases. Under acute and chronic stress, the number of immune cells increases and therefore the rate of cell division in immune cells rises. This increased rate of cell division results in a faster loss of telomere sequences. Moreover, cell culture studies have demonstrated that cortisol concentrations that are equal to physiological conditions suppress telomerase activity, which additionally leads to further telomere shortening.[76] This highlights the importance of an interdisciplinary, multiperspective approach in studying the mechanisms and consequences of CM and TCM, respectively. Joining forces in the field of CM and investigating the

psychological mechanisms, risk, and protective factors in CM, and relating these to the various biological systems, is necessary in future studies. In particular, longitudinal designs that investigate psychological and biological functioning at the same time are necessary to gain a deeper understanding of the mechanisms in CM and TCM and, finally, to develop tailored early interventions for families and children at risk.

Acknowledgments

We thank Sarah Wilker and Alexander Karabatsiakis for comments on earlier versions of the manuscript.

Conflicts of interest

The authors declare no conflicts of interest.

References

1. Briere, J. 1992. *Child Abuse Trauma: Theory and Treatment of the Lasting Effects*. SAGE. Newbury Park, CA.
2. Cicchetti, D. & S.L. Toth. 2005. Child maltreatment. *Annu. Rev. Clin. Psychol.* **1:** 409–438.
3. Myers, J.E.B. *et al.* 2002. *APSAC Handbook on Child Maltreatment*. SAGE. Thousand Oaks, CA.

4. Dong, M. *et al.* 2004. The interrelatedness of multiple forms of childhood abuse, neglect, and household dysfunction. *Child Abuse Negl.* **28:** 771–784.

5. Green, J.G. *et al.* 2010. Childhood adversities and adult psychiatric disorders in the national comorbidity survey replication I: associations with first onset of DSM-IV disorders. *Arch. Gen. Psychiatry* **67:** 113–123.

6. Kolassa, I.-T. *et al.* 2010a. The risk of posttraumatic stress disorder after trauma depends on traumatic load and the catechol-o-methyltransferase Val(158)Met polymorphism. *Biol. Psychiatry* **67:** 304–308.

7. Kolassa, I.-T. *et al.* 2010b. Spontaneous remission from PTSD depends on the number of traumatic event types experienced. *Psychol. Trauma* **2:** 169–174.

8. Neuner, F. *et al.* 2004. Psychological trauma and evidence for enhanced vulnerability for posttraumatic stress disorder through previous trauma among West Nile refugees. *BMC Psychiatr.* **4:** 34.

9. Anda, R.F. *et al.* 2006. The enduring effects of abuse and related adverse experiences in childhood. A convergence of evidence from neurobiology and epidemiology. *Eur. Arch. Psy. Clin.* **256:** 174–186.

10. Duke, N.N. *et al.* 2010. Adolescent violence perpetration: associations with multiple types of adverse childhood experiences. *Pediatrics* **125:** e778–e786.

11. Gilbert, R. *et al.* 2009. Burden and consequences of child maltreatment in high-income countries. *Lancet* **373:** 68–81.

12. Briere, J. & C.E. Jordan. 2009. Childhood maltreatment, intervening variables, and adult psychological difficulties in women: an overview. *Trauma Violence Abuse* **10:** 375–388.

13. Anda, R.F. *et al.* 2010. Adverse childhood experiences and frequent headaches in adults. *Headache* **50:** 1473–1481.

14. Brown, D.W. *et al.* 2009. Adverse childhood experiences and the risk of premature mortality. *Am. J. Prev. Med.* **37:** 389–396.

15. Felitti, V.J. *et al.* 1998. Relationship of childhood abuse and household dysfunction to many of the leading causes of death in adults. The Adverse Childhood Experiences (ACE) Study. *Am. J. Prev. Med.* **14:** 245–258.

16. Panzer, A. 2008. The neuroendocrinological sequelae of stress during brain development: the impact of child abuse and neglect. *Afr. J. Psychiatr.* **11:** 29–34.

17. Dixon, L. *et al.* 2009. Patterns of risk and protective factors in the intergenerational cycle of maltreatment. *J. Fam. Violence* **24:** 111–122.

18. Pears, K.C. & D.M. Capaldi. 2001. Intergenerational transmission of abuse: a two-generational prospective study of an at-risk sample. *Child Abuse Neglect* **25:** 1439–1461.

19. Sidebotham, P. & J. Golding. 2001. Child maltreatment in the "children of the nineties" a longitudinal study of parental risk factors. *Child Abuse Neglect* **25:** 1177–1200.

20. Berlin, L.J. *et al.* 2011. Intergenerational continuity in child maltreatment: mediating mechanisms and implications for prevention. *Child Dev.* **82:** 162–176.

21. Browne, K.D. & M. Herbert. 1997. *Preventing Family Violence.* Wiley. Chichester, UK.

22. Kaufmann, J. & E. Ziegler. 1989. *The Intergenerational Transmission of Child Abuse. Child Maltreatment: Theory and Re-search on the Causes and Consequences of Child Abuse and Neglect,* pp. 129–150. Cambridge University Press. Cambridge.

23. Dixon, L. *et al.* 2005. Risk factors of parents abused as children: a mediational analysis of the intergenerational continuity of child maltreatment (Part I). *J. Child Psychol. Psychiatry* **46:** 47–57.

24. Kaufmann, J. & C. Henrich. 2000. Exposure to violence and early childhood trauma. In *Handbook of Infant Mental Health,* pp. 195–207. Guilford Press. New York.

25. Hunter, A.L. *et al.* 2011. Altered stress responses in children exposed to early adversity: a systematic review of salivary cortisol studies. *Stress Early Online* 1–3.

26. Holsboer, F. 2001. Stress, hypercortisolism and corticosteroid receptors in depression: implications for therapy. *J. Affect Dosorders* **62:** 77–91.

27. Meewisse, M.-L. *et al.* 2007. Cortisol and post-traumatic stress disorder in adults: systematic review and meta-analysis. *Br. J. Psychiatry* **191:** 387–392.

28. Heim, C. *et al.* 2000. The potential role of hypocortisolism in the pathophysiology of stress-related bodily disorders. *Psychoneuroendocrino* **25:** 1–35.

29. Cicchetti, D. & F.A. Rogosch. 2001. The impact of child maltreatment and psychopathology on neuroendocrine functioning. *Dev. Psychopathol.* **13:** 783–804.

30. Gow, R. *et al.* 2010. An assessment of cortisol analysis in hair and its clinical applications. *Forensic Sci. Int.* **196:** 32–37.

31. Guilliams, T.G. & L. Edwards. 2010. Chronic stress and the HPA axis: clinical assessment and therapeutic considerations. *Standard* **9:** 1–12.

32. Fisher, D.B. *et al.* 2007. Intergenerational predictors of diurnal cortisol secretion in early childhood. *Infant Child Dev.* **170:** 151–170.

33. Brand, S.R. *et al.* 2010. The impact of maternal childhood abuse on maternal and infant HPA axis function in the postpartum period. *Psychoneuroendocrino* **35:** 686–693.

34. Brennan, P.A. *et al.* 2008. Maternal depression and infant cortisol: influences of timing, comorbidity and treatment. *J. Child Psychol. Psychiatry* **49:** 1099–1107.

35. Laurent, H.K. *et al.* 2011. Risky shifts: how the timing and course of mothers' depressive symptoms across the perinatal period shape their own and infant's stress response profiles. *Dev. Psychopathol.* **23:** 521–538.

36. Yehuda, R. *et al.* 2007. Parental posttraumatic stress disorder as a vulnerability factor for low cortisol trait in offspring of holocaust survivors. *Arch. Gen. Psychiatry* **64:** 1040–1048.

37. Yehuda, R. *et al.* 2005. Transgenerational effects of posttraumatic stress disorder in babies of mothers exposed to the World Trade Center attacks during pregnancy. *J. Clin. Endocrinol. Metab.* **90:** 4115–4118.

38. Epel, E.S. *et al.* 2004. Accelerated telomere shortening in response to life stress. *Proc. Natl. Sci. U.S.A.* **101:** 17312–17315.

39. Chan, S.R.W.L. & E.H. Blackburn. 2004. Telomeres and telomerase. *Philos. Trans. R. Soc. London, Ser. B* **359:** 109–121.

40. Jiang, H. *et al.* 2008. Proteins induced by telomere dysfunction and DNA damage represent biomarkers of human aging and disease. *Proc. Natl. Acad. Sci. U.S.A.* **105:** 11299–11304.

41. Chan, S.R.W.L. & E.H. Blackburn. 2003. Telomerase and ATM/Tel1p protect telomeres from nonhomologous end joining. *Molecular Cell* **11**: 1379–1387.

42. Fitzpatrick, A.L. *et al.* 2007. Leukocyte telomere length and cardiovascular disease in the cardiovascular health study. *Am. J. Epidemiol.* **165**: 14–21.

43. Willeit, P. *et al.* 2010. Telomere length and risk of incident cancer and cancer mortality. *JAMA* **304**: 69–75.

44. Cawthon, R.M. *et al.* 2003. Association between telomere length in blood and mortality in people aged 60 years or older. *Lancet* **361**: 393–395.

45. Tyrka, A.R. *et al.* 2010. Childhood maltreatment and telomere shortening: preliminary support for an effect of early stress on cellular aging. *Biol. Psychiatry* **67**: 531–534.

46. Kiecolt-Glaser, J.K. *et al.* 2010. Childhood adversity heightens the impact of later-life caregiving stress on telomere length and inflammation. *Psychosom. Med.* **73**: 16–22.

47. Kananen, L. *et al.* 2010. Childhood adversities are associated with shorter telomere length at adult age both in individuals with an anxiety disorder and controls. *PLoS ONE* **5**: e10826.

48. Drury, S.S. *et al.* 2012. Telomere length and early severe social deprivation: linking early adversity and cellular aging. *Mol. Psychiatry* In press.

49. O'Donovan, A. *et al.* 2011. Childhood trauma associated with short leukocyte telomere length in posttraumatic stress disorder. *Biol. Psychiatry* **70**: 465–471.

50. Glass, D. *et al.* 2010. No correlation between childhood maltreatment and telomere length. *Biol. Psychiatry* **68**: e21–e22.

51. Dantzer, R. *et al.* 2008. Frim inflammation to sickness and depression: when the immune system subjugates the brain. *Nat. Rev. Neurosci.* **9**: 46–56.

52. McGeer, P.L. & E.G. McGeer. 2004. Inflammation and the degenerative diseases of aging. *Ann. N.Y. Acad. Sci.* **1035**: 104–116.

53. Yudkin, J.S. *et al.* 2000. Inflammation, obesity, stress and coronary heart disease: is interleukin-6 the link? *Atherosclerosis* **148**: 209–214.

54. Danese, A. *et al.* 2009. Adverse childhood experiences and adult risk factors for age-related disease: depression, inflammation, and clustering of metabolic risk markers. *Arch. Pediatr. Med.* **163**: 1135–1143.

55. Danese, A. 2007. Childhood maltreatment predicts adult inflammation in a life-course study. *Proc. Natl. Acad. Sci. U.S.A.* **104**: 1319–1324.

56. Carpenter, L.L. *et al.* 2010. Association between plasma IL-6 response to acute stress and early-life adversity in healthy adults. *Neuropsychopharmacology* **35**: 2617–2623.

57. Sternthal, M.J. *et al.* 2009. Maternal interpersonal trauma and cord blood IgE levels in an inner-city cohort: a life-course perspective. *J. Allergy Clin. Immunol.* **124**: 954–960.

58. Halken, S. 2003. Early sensitisation and development of allergic airway disease—risk factors and predictors. *Paediatr. Respir. Rev.* **4**: 128–134.

59. Lemke, H. *et al.* 2009. Benefits and burden of the maternally-mediated immunological imprinting. *Autoimmun. Rev.* **8**: 394–399.

60. Wright, R.J. 2007. Prenatal maternal stress and early caregiving experiences: implications for childhood asthma risk. *Paediatr. Perinatal. Epidemiol.* **21**: 8–14.

61. Peden, D.B. 2000. Development of atopy and asthma: candidate environmental influences and important periods of exposure. *Environ. Health Perspect.* **108**: 475–482.

62. Champagne, F. A. 2008. Epigenetic mechanisms and the transgenerational effects of maternal care. *Front. Neuroendocrinol.* **29**: 386–397.

63. Arai, J.A. & L.A. Feig. 2010. Long-lasting and transgenerational effects of an environmental enrichment on memory formation. *Brain Res. Bull.* **85**: 30–35.

64. Turner, B. 2001. *Chromatin and Gene Regulation.* Blackwell Science. Oxford.

65. Radtke, K.M. *et al.* 2011. Transgenerational impact of intimate partner violence on methylation in the promoter of the glucocorticoid receptor. *Transl. Psychiatry* **1**: e21.

66. Weaver, I.C.G. *et al.* 2004. Epigenetic programming by maternal behavior. *Nat. Neurosci.* **7**: 847–854.

67. Zhang, T.-Y. *et al.* 2010. Maternal care and DNA methylation of a glutamic acid decarboxylase 1 promoter in rat hippocampus. *J. Neurosci.* **30**: 13130–13137.

68. Roth, T.L. *et al.* 2009. Lasting epigenetic influence of early-life adversity on the BDNF gene. *Biol. Psychiatry* **65**: 760–769.

69. Daniels, W. *et al.* 2009. Maternal separation alters nerve growth factor and corticosterone levels but not the DNA methylation status of the exon 1 7 glucocorticoid receptor promoter region. *Metab. Brain Dis.* **24**: 615–627.

70. Franklin, T.B. & I.M. Mansuy. 2010. Epigenetic inheritance in mammals: evidence for the impact of adverse environmental effects. *Neurobiol. Dis.* **39**: 61–65.

71. Lange, U.C. & R. Schneider. 2010. What an epigenome remembers. *Bioessays* **32**: 659–668.

72. Oberlander, T.F. *et al.* 2008. Prenatal exposure to maternal depression, neonatal methylation of human glucocorticoid receptor gene (NR3C1) and infant cortisol stress responses. *Epigenetics* **3**: 97–106.

73. Devlin, A.M. *et al.* 2010. Prenatal exposure to maternal depressed mood and the MTHFR C677T variant affect SLC6A4 methylation in infants at birth. *PloS ONE* **5**: e12201.

74. McGowan, P.O. *et al.* 2009. Epigenetic regulation of the glucocorticoid receptor in human brain associates with childhood abuse. *Nat. Neurosci.* **12**: 342–348.

75. Dhabhar, F.S. & B.S. McEwen. 2001. Bidirectional effects of stress and glucocorticoid hormones on immune function: possible explanations for paradoxical observations. In *Psychoneuroimmunology,* 3rd Ed. R. Ader, D.L. Felton & N. Cohen, Eds.: 301–338. Academic Press. New York.

76. Choi, J. *et al.* 2008. Reduced telomerase activity in human T lymphocytes exposed to cortisol. *Brain Behav. Immun.* **22**: 600–605.

Ann. N.Y. Acad. Sci. ISSN 0077-8923

ANNALS OF THE NEW YORK ACADEMY OF SCIENCES
Issue: *Neuroimmunomodulation in Health and Disease*

Chronic neuropathic pain–like behavior and brain-borne IL-1β

Adriana del Rey,[1] A. Vania Apkarian,[2] Marco Martina,[2] and Hugo O. Besedovsky[1]

[1]Division of Immunophysiology, Institute of Physiology and Pathophysiology, Marburg, Germany. [2]Department of Physiology, Northwestern University, Feinberg School of Medicine, Chicago, Illinois

Address for correspondence: Adriana del Rey, Division of Immunophysiology, Institute of Physiology and Pathophysiology, Deutschhausstrasse 2, 35037 Marburg, Germany. delrey@mailer.uni-marburg.de

Neuropathic pain in animals results in increased IL-1β expression in the damaged nerve, the dorsal root ganglia, and the spinal cord. Here, we discuss our results showing that this cytokine is also overexpressed at supraspinal brain regions, in particular in the contralateral side of the hippocampus and prefrontal cortex and in the brainstem, in rats with neuropathic pain-like behavior. We show that neuropathic pain degree and development depend on the specific nerve injury model and rat strain studied, and that there is a correlation between hippocampal IL-1β expression and tactile sensitivity. Furthermore, the correlations between hippocampal IL-1β and IL-1ra or IL-6 observed in control animals, are disrupted in rats with increased pain sensitivity. The lateralization of increased cytokine expression indicates that this alteration may reflect nociception. The potential functional consequences of increased IL-1β expression in the brain during neuropathic pain are discussed.

Keywords: chronic pain; interleukin-1; hippocampus; lateralization

Introduction

According to the American Academy of Pain Medicine,[1] seven- and fourfold more Americans suffer from chronic pain than from stroke or diabetes, respectively. Not less impressive are the statistics provided by a recent research report,[2] which estimates that more than 1.5 billion people worldwide suffer from chronic pain, and that approximately 3–4.5% of the global population are afflicted with neuropathic pain, resulting in concerted efforts by the scientific community to tackle this health problem. The approaches range from the identification of the molecules that might be involved in the persistence of pain to the emotional experience associated with it. Preceded by a few short general comments, we refer here to recent studies, mainly derived from our laboratories, focusing on a very specific question: namely, whether brain-borne IL-1β could be one of the factors involved in neuropathic chronic pain, an aspect that, at present, can only be explored in animal models.

Neuropathic pain in animal models

Neuropathic pain is a persistent type of pain that usually results from injury or disease of the nervous system. Neuropathic pain was originally classified as central or peripheral depending on whether the primary insult is on the peripheral or the central nervous system (CNS). In humans, central neuropathic pain is found, for example, in spinal cord injury and multiple sclerosis, while common causes of peripheral neuropathic pain are infection with herpes zoster, nutritional deficiencies, compression of a peripheral nerve by a tumor, or physical trauma to a nerve trunk. This separation, however, does not seem to be tenable at present because of the pathophysiological mechanisms involved. A discussion of these complex mechanisms is beyond the scope of this paper, and the reader is referred to excellent reviews on this topic.[3–5]

The most common models to study neuropathic pain in animals are based on manipulations of a peripheral nerve, either by a chronic constriction injury (CCI) or a spare nerve injury (SNI). In the CCI

doi: 10.1111/j.1749-6632.2012.06621.x
Ann. N.Y. Acad. Sci. 1262 (2012) 101–107 © 2012 New York Academy of Sciences.

model, one sciatic nerve is exposed above the trifurcation and ligated with knots. In the SNI model, one sciatic nerve is exposed at the level of its trifurcation, each of the tibial and common peroneal nerves are completely severed in between, and the sural nerve is left intact. It has been reported that SNI results in irreversible damage of the sciatic nerve, whereas the effects of CCI are at least in part reversible.[6,7] Hyperalgesia, which implies an extreme reaction to a moderate stimulus, and allodynia, a withdrawal response to a nonnoxious tactile or thermal stimulus that normally does not provoke pain, are among the behavioral parameters that are evaluated as indication of pain in these models. In most studies of this type, the response observed at the side contralateral to that in which the nerve was injured or sham-operated animals are used as controls.

IL-1β can induce pain

A review of the literature makes it difficult to fix the historical point of not only when the first cytokine was discovered but also which one was the first discovered. Cytokines seem to have been rediscovered, with new functions being reported in the last years, depending on the particular field of research. It is likely that at least one of the first cytokines discovered was IL-1.[8] In relation to pain, Ferreira *et al.* showed more than 20 years ago that IL-1β can act as a potent hyperalgesic agent.[9] In these experiments, IL-1β was injected intraplantarally or intraperitoneally, and the authors proposed that the cytokine acts at peripheral levels. This result has been confirmed by other groups[10] (recent reviews and references on how cytokines can induce pain acting at peripheral levels can be found in Refs. 11 and 12). However, more recently, it has also been shown that, although animals that lack IL-1β signaling show reduced mechanical allodynia, the cytokine is necessary to recover sciatic nerve functions.[13]

Other cytokines that have been consistently demonstrated to be involved in neuropathic pain are TNF-α and IL-6 (for review, see Refs. 3 and 14).

Peripheral nerve injury induces IL-1β expression at local and spinal cord levels

Sciatic nerve transection and CCI result in increased local IL-1 expression in the damaged nerve, the dorsal root ganglia, and the spinal cord (see Refs. 15 and 19, and for review, see Refs. 12 and 14). One intriguing fact is that some groups have reported that IL-1β is also expressed in the contralateral, nonoperated rat sciatic nerve.[20,21]

In a model of sciatic nerve microcrush lesions, it has been shown that IL-1 expression and recovery of locomotor function is MyD88 dependent.[22] This finding is interesting in this context because this adaptor protein mediates, although not exclusively, the known proinflammatory effects of IL-1.

IL-1β expression at supraspinal levels during chronic pain

There is ample evidence that neuropathic pain–behavior in rodents is accompanied by reorganization of peripheral and spinal cord nociceptive processing, giving rise to central sensitization.[23] However, cytokine expression in pain states has rarely been studied above the spinal cord. We started our studies on the possible expression of IL-1β in selected brain regions because of evidence showing that this cytokine is involved in the modulation of nociceptive information,[24] and investigation of the hypothesis that supraspinal circuitry plays a role in the transition from acute to chronic pain.[25,26] Restricting the discussion to the CNS, the pharmacological studies available indicated that central administration of IL-1β can induce analgesia or hyperalgesia depending on the brain region involved and on the dose injected: lower doses cause hyperalgesia and higher doses are analgesic.[24,27] However, at that time, there were no reports on possible changes in endogenous IL-1β expression at supraspinal levels during chronic neuropathic pain.

We first examined IL-1β mRNA expression in the brainstem, thalamus, and prefrontal cortex in the two models of neuropathic pain mentioned earlier, 10 and 24 days after nerve injury, in male Wistar–Kyoto rats.[28] One reason to choose these regions is the anatomy of nociceptive transmission. The thalamus is the main termination site for the spinothalamic pathway, and the brainstem contains structures for ascending nociceptive pathways and important descending modulatory structures. Another reason is that metabolic abnormalities and decreased gray matter density have been reported in the prefrontal cortex and thalamus of patients with chronic back pain.[29,30] Neuropathic pain–like behavior was monitored by tracking tactile thresholds. Neither significant changes in threshold sensitivity nor in IL-1β expression in the brain regions mentioned were detected in CCI-injured rats. In contrast, SNI-injured animals showed a robust and sustained decrease in tactile thresholds of the injured foot, and increased IL-1β expression was observed in the brainstem and in the prefrontal cortex contralateral to the injury side 10 days after injury. Only a modest increase in IL-1β expression was observed in the thalamus/striatum in the side contralateral to the injury on day 24. Interestingly, a positive correlation between threshold sensitivity in the injured paw and IL-1 expression on the ipsilateral side of the prefrontal cortex was detected in SNI-rats, whereas no correlation was present in sham-operated or untouched control animals. These results were the first indication that neuroimmune activation in neuropathic pain conditions includes supraspinal brain regions, and suggested that IL-1β modulation of supraspinal circuitry of pain may be involved in neuropathic conditions.[28]

We then explored whether IL-1β expression was altered also in the hippocampus during chronic neuropathic pain. Several reasons motivated us to include this region. The first was that the hippocampus has been proposed as one of the critical limbic components involved in the transition from acute to chronic pain.[25,26] Second, the hippocampus interacts with the prefrontal cortex, a region in which, as mentioned earlier, we had detected increased IL-1β expression during neuropathic pain, and that therefore may be correlated with a potential expression of this cytokine in the hippocampus. Furthermore, there was evidence of abnormal hippocampal function (for review, see Ref. 31) and impaired neurogenesis[32,33] in chronic pain. Another line of evidence, related to long-term potentiation (LTP) and learning, has indicated that the hippocampus may be a potential region in which IL-1β could be altered during neuropathic pain. A recent study found that neuropathic animals exhibit working memory deficits and decreased LTP, although the changes were not related to the pain behavior of the animals.[34] On the other hand, we have shown that *in vivo* and *in vitro* LTP induction in the hippocampus results in a long-lasting increase in IL-1β expression, and that blockade of IL-1 signaling impairs the maintenance of LTP.[35] It has also been shown that IL-1 can affect learning of hippocampus-dependent tasks.[36–38] Finally, another group showed that the expression of IL-1β is decreased in the hippocampus hours after peripheral nerve injury.[39] In our view, however, the extent to which the changes described

reflected general stress or effects of pain remained unclear. Furthermore, our main purpose was to explore alterations in IL-1 expression that could occur during the more chronic phase of neuropathic pain, and not early immediate effects that could be evoked soon after nerve injury. In any case, taken together, the evidence summarized indicated the possibility that IL-1β expression in the hippocampus could be affected during chronic neuropathic pain. To address this hypothesis, we chose again the two models of neuropathic pain mentioned earlier because they display different degrees of pain behavior in response to nerve injury.[6,40] We also included two different rat strains (Wistar–Kyoto and Sprague–Dawley) because it is known that there is a large variation in neuropathic pain behavior not only between inbred and outbred strains, but also even between substrains.[41,42] On the basis of the results showing changes in IL-1β gene expression in the prefrontal cortex contralateral to the injury side, we first concentrated on the evaluation of this cytokine in the contralateral side of the hippocampus at different times after nerve injury. We found that in fact both injury models resulted in pain-like behaviors that differed in intensity and duration in the two rat strains, and the changes in IL-1β gene expression in the hippocampus agreed with these differences.[43]

CCI, which did not produce any significant effect on tactile thresholds in Wistar–Kyoto rats, failed to induce significant changes in hippocampal IL-1β expression compared with the sham-operated controls. However, SNI resulted in elevated mechanical hypersensitivity in this strain, and IL-1β mRNA levels were significantly increased 10 and 24 days after injury in the contralateral hippocampus. In Sprague–Dawley rats both injury models resulted in significantly decreased tactile thresholds. Interestingly, while the allodynia was maintained for more than 2 months in the animals that had undergone SNI, the decreased threshold sensitivity following CCI was nearly recovered by this time. In line with these behavioral effects, IL-1β mRNA expression in the contralateral hippocampus was significantly increased 2 weeks after surgery in both pain models, but remained significantly elevated up to 2 months only in rats with SNI. These results are schematically summarized in Figure 1.

In addition, in the hippocampus, the changes were restricted to the contralateral hippocampus, and no significant changes were detected in the side of the hippocampus ipsilateral to the nerve injury.

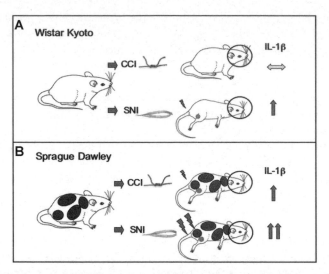

Figure 1. IL-1β expression in the contralateral hippocampus depends on the pain model and rat strain. (A) In Wistar–Kyoto rats, while CCI did not produce any significant effect on tactile thresholds and no significant changes in hippocampal IL-1β expression, SNI resulted in elevated mechanical hypersensitivity and significantly increased IL-1β mRNA levels in the contralateral hippocampal side. (B) In Sprague–Dawley rats, both injury models resulted in significantly decreased tactile thresholds. CCI-induced decreased threshold sensititvity recovered gradually, whereas allodynia was maintained for more than 2 months in the animals that had undergone SNI. The temporal expression on IL-1β mRNA in the contralateral hippocampus agreed with these behavioral effects.

This clear hemispheric lateralization of cytokine expression following SNI or CCI indicates that the changes in cytokine expression are not due to a stress response, because the latter is expected to result in bilateral effects. Our results apparently contradict those reported by Al-Amin *et al.*,[44] who reported bilateral increases in hippocampal IL-1β levels and even more expression in the ipsilateral than in the contralateral side in the CCI model. However, differences in the experimental design (e.g., in the mentioned publication, the animals received daily intraperitoneal injections of a vehicle), animal sex, and the intensity of the pain manifestation could account for this discrepancy.

Supporting a link between the pain-like behavior, as reflected by the tactile threshold, and IL-1β overexpression in the contralateral hippocampus, a linear correlation between both variables was observed in both strains and for both nerve injury models. Allodynia, when present, was only manifested in the hind paw ipsilateral to the peripheral nerve injury.

Another interesting observation that derived from these studies was that the effect of SNI and CCI on the endogenous IL-1 receptor antagonist (IL-1ra) and IL-6 expression in the hippocampus was different in Sprague–Dawley as compared with

that in Wistar–Kyoto rats. IL-6 was only increased in Wistar–Kyoto rats with SNI 10 days after surgery, and no significant changes were detected for this cytokine in Sprague–Dawley rats in any of the two models. No significant changes in IL-1ra were observed in Wistar–Kyoto rats with CCI or SNI, but the IL-1 antagonist was significantly increased in Sprague–Dawley rats in both injury models (for details about these results, see Ref. 43). Interestingly, IL-1β mRNA levels in the hippocampus of sham-operated Sprague–Dawley rats positively correlated with those of IL-1ra and IL-6. The fact that these correlations were not observed in animals with peripheral nerve injury indicates that the interactions between these cytokine are disrupted in the hippocampus of animals with chronic pain.

In summary, this study showed that (1) the degree and development of neuropathic pain are correlated with specific nerve injury models and rat strains; (2) hippocampal IL-1β expression is correlated with pain behavior; (3) the alterations in cytokine expression are restricted to brain regions contralateral to the injury side; and (4) neuropathic pain seems to disrupt a network of cytokine interactions in the hippocampus. These results are schematically shown in Figure 2.

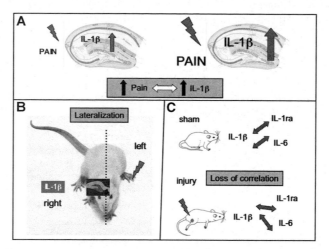

Figure 2. IL-1β expression in the contralateral hippocampus correlates with the intensity of pain-like behavior, but not with IL-6 or IL-1ra. (A) IL-1β overexpression in the hippocampus correlates with pain-like behavior, as reflected by the tactile thresholds, in both strains and for both nerve injury models. (B) The changes in IL-1β expression are restricted to the contralateral side of the hippocampus. (C) Although IL-1β expression positively corrrelated with IL-6 and IL-1ra expression in the contralateral hippocampus of sham-operated rats, these correlations are lost in animals in which the left sciatic nerve was injured.

The different intensity and duration of the nerve injuries in the pain-like behavior of Wistar–Kyoto and Sprague–Dawley rats proved to be a good choice for our purposes. At the same time, the results highlight the importance of considering not only the species, but also the strain and model of injury when studying chronic pain. More importantly, however, they point to the difficulty of extrapolating results from animal experiments to humans with chronic neuropathic pain.

What are the functional consequences of increased IL-1β expression in the brain during neuropathic pain?

The findings discussed earlier indicate a causal relation between neuropathic pain–like behavior and expression of IL-1β in the brain regions studied, implying synaptic reorganization of these regions. However, a positive relation between threshold sensitivity and brain-borne cytokine expression does not allow for concluding whether the animals showed increased allodynia because IL-1β was increased, or, conversely, if increased cytokine expression is the result of increased pain-like behavior. Studies showing that pain behavior is attenuated in animals in which IL-1 signaling is permanently impeded[45,46] cannot answer this question. We are aware of only one publication using neuropathic

pain models in which IL-1 was neutralized directly in the brain and where the functional consequence of IL-1 neutralization on mechanical allodynia was tested. In this study, injection of an anti-IL-1 antibody in the red nucleus significantly attenuated mechanical allodynia, but this was not paralleled by changes in IL-1β immunoreactivity in the hippocampus.[47] Studies in which IL-1 was neutralized by peripheral or intrathecal administration of IL-1ra[45,48] found that this treatment attenuates the allodynic response to peripheral nerve injury. However, direct proof that brain-borne IL-1 plays a functional role during pain could only be obtained by blocking its effects in defined brain regions in which the cytokine is overexpressed, and at defined times after injury. In addition to the intrinsic experimental difficulties, this would also affect several neuroendocrine, metabolic, and immune parameters (for review, see Ref. 49).

As mentioned, IL-1β is involved in the regulation of synaptic plasticity and memory processes.[35–37,50–52] Thus, it is conceivable that changes in IL-1 expression in the brain would have functional consequences for the deficits in cognitive functions reported in patients and animals with chronic pain.[34,53–55] However, we are not aware of any study at present that has directly approached these possible links. Recently published evidence that acute interference of IL-1 signals in the brain reduces the depressive-like symptoms observed during SNI[56] indicates that overexpression of IL-1 in the hippocampus may be relevant for these symptoms.

It is clear that much more needs to be done to clarify the pathophysiological links among brain-borne IL-1, chronic neuropathic pain, and brain functions. Adding to this already complicated scenario, it seems necessary to stress again that we have concentrated here on discussing only IL-1β; however, this cytokine is part of a complex network composed of other cytokines as well, such as IL-6 and TNF-α, which are known to be also involved in neuropathic pain behavior.

Acknowledgments

The studies described in this paper were funded by NIH NINDS RO1 NS057704 (A.V.A.), and NINDS RO1 NS064091 (M.M.).

Conflicts of interest

The authors declare no conflicts of interest.

References

1. American Academy of Pain Medicine. 2012. URL: http://www.painmed.org/patientcenter/facts_on _pain.aspx [Accessed on 1 June 2012]
2. Global Industry Analysts. 2011. URL: http://www.prweb.com/releases/2011/1/prweb8052240.htm [Accessed on 1 June 2012].
3. Myers, R.R., W.M. Campana & V.I. Shubayev. 2006. The role of neuroinflammation in neuropathic pain: mechanisms and therapeutic targets. *Drug Discov. Today* **11:** 8–20.
4. Pace, M.C. *et al.* 2006. Neurobiology of pain. *J. Cell. Physiol.* **209:** 8–12.
5. Zimmermann, M. 2001. Pathobiology of neuropathic pain. *Eur. J. Pharmacol.* **429:** 23–37.
6. Bennett, G.J. & Y.K. Xie. 1988. A peripheral mononeuropathy in rat that produces disorders of pain sensation like those seen in man. *Pain* **33:** 87–107.
7. Kim, K.J., Y.W. Yoon & J.M. Chung. 1997. Comparison of three rodent neuropathic pain models. *Exp. Brain Res.* **113:** 200–206.
8. Dinarello, C.A. 2011. A clinical perspective of IL-1beta as the gatekeeper of inflammation. *Eur. J. Immunol.* **41:** 1203–1217.
9. Ferreira, S.H. *et al.* 1988. Interleukin-1 beta as a potent hyperalgesic agent antagonized by a tripeptide analogue. *Nature* **334:** 698–700.
10. Zelenka, M., M. Schafers & C. Sommer. 2005. Intraneural injection of interleukin-1beta and tumor necrosis factor-alpha into rat sciatic nerve at physiological doses induces signs of neuropathic pain. *Pain* **116:** 257–263.
11. Sommer, C. & M. Kress. 2004. Recent findings on how proinflammatory cytokines cause pain: peripheral mechanisms in

12. Ren, K. & R. Torres. 2009. Role of interleukin-1beta during pain and inflammation. *Brain Res. Rev.* **60:** 57–64.
13. Nadeau, S. *et al.* 2011. Functional recovery after peripheral nerve injury is dependent on the *proinflammatory* cytokines IL-1beta and TNF: implications for neuropathic pain. *J. Neurosci.* **31:** 12533–12542.
14. Uceyler, N. & C. Sommer. 2008. Cytokine regulation in animal models of neuropathic pain and in human diseases. *Neurosci. Lett.* **437:** 194–198.
15. DeLeo, J.A., R.W. Colburn & A.J. Rickman. 1997. Cytokine and growth factor immunohistochemical spinal profiles in two animal models of mononeuropathy. *Brain Res.* **759:** 50–57.
16. Shamash, S., F. Reichert & S. Rotshenker. 2002. The cytokine network of Wallerian degeneration: tumor necrosis factor-alpha, interleukin-1alpha, and interleukin-1beta. *J. Neurosci.* **22:** 3052–3060.
17. Lee, H.L. *et al.* 2004. Temporal expression of cytokines and their receptors mRNAs in a neuropathic pain model. *Neuroreport* **15:** 2807–2811.
18. Rotshenker, S., S. Aamar & V. Barak. 1992. Interleukin-1 activity in lesioned peripheral nerve. *J. Neuroimmunol.* **39:** 75–80.
19. Okamoto, K. *et al.* 2001. Pro- and anti-inflammatory cytokine gene expression in rat sciatic nerve chronic constriction injury model of neuropathic pain. *Exp. Neurol.* **169:** 386–391.
20. Kleinschnitz, C. *et al.* 2005. Contralateral cytokine gene induction after peripheral nerve lesions: dependence on the mode of injury and NMDA receptor signaling. *Brain Res. Mol. Brain Res.* **136:** 23–28.
21. Ruohonen, S. *et al.* 2002. Contralateral non-operated nerve to transected rat sciatic nerve shows increased expression of IL-1beta, TGF-beta1, TNF-alpha, and IL-10. *J. Neuroimmunol.* **132:** 11–17.
22. Boivin, A. *et al.* 2007. Toll-like receptor signaling is critical for Wallerian degeneration and functional recovery after peripheral nerve injury. *J. Neurosci.* **27:** 12565–12576.
23. Woolf, C.J. & M.W. Salter. 2000. Neuronal plasticity: increasing the gain in pain. *Science* **288:** 1765–1769.
24. Bianchi, M., B. Dib & A.E. Panerai. 1998. Interleukin-1 and nociception in the rat. *J. Neurosci. Res.* **53:** 645–650.
25. Apkarian, A.V. 2008. Pain perception in relation to emotional learning. *Curr. Opin. Neurobiol.* **18:** 464–468.
26. Apkarian, A.V., J.A. Hashmi & M.N. Baliki. 2011. Pain and the brain: specificity and plasticity of the brain in clinical chronic pain. *Pain* **152:** S49–S64.
27. Hori, T. *et al.* 1998. Pain modulatory actions of cytokines and prostaglandin E2 in the brain. *Ann. N.Y. Acad. Sci.* **840:** 269–281.
28. Apkarian, A.V. *et al.* 2006. Expression of IL-1beta in supraspinal brain regions in rats with neuropathic pain. *Neurosci. Lett.* **407:** 176–181.
29. Apkarian, A.V. *et al.* 2004. Chronic back pain is associated with decreased prefrontal and thalamic gray matter density. *J. Neurosci.* **24:** 10410–10415.

30. Grachev, I.D., B.E. Fredrickson & A.V. Apkarian. 2000. Abnormal brain chemistry in chronic back pain: an in vivo proton magnetic resonance spectroscopy study. *Pain* **89:** 7–18.

31. Liu, M.G. & J. Chen. 2009. Roles of the hippocampal formation in pain information processing. *Neurosci. Bull.* **25:** 237–266.

32. Duric, V. & K.E. McCarson. 2006. Persistent pain produces stress-like alterations in hippocampal neurogenesis and gene expression. *J. Pain* **7:** 544–555.

33. Terada, M. *et al.* 2008. Suppression of enriched environment-induced neurogenesis in a rodent model of neuropathic pain. *Neurosci. Lett.* **440:** 314–318.

34. Ren, W.J. *et al.* 2011. Peripheral nerve injury leads to working memory deficits and dysfunction of the hippocampus by upregulation of TNF-alpha in rodents. *Neuropsychopharmacology* **36:** 979–992.

35. Schneider, H. *et al.* 1998. A neuromodulatory role of interleukin-1beta in the hippocampus. *Proc. Natl. Acad. Sci. U.S.A.* **95:** 7778–7783.

36. Avital, A. *et al.* 2003. Impaired interleukin-1 signaling is associated with deficits in hippocampal memory processes and neural plasticity. *Hippocampus* **13:** 826–834.

37. Depino, A.M. *et al.* 2004. Learning modulation by endogenous hippocampal IL-1: blockade of endogenous IL-1 facilitates memory formation. *Hippocampus* **14:** 526–535.

38. Yirmiya, R., G. Winocur & I. Goshen. 2002. Brain interleukin-1 is involved in spatial memory and passive avoidance conditioning. *Neurobiol. Learn. Mem.* **78:** 379–389.

39. Uceyler, N., A. Tscharke & C. Sommer. 2008. Early cytokine gene expression in mouse CNS after peripheral nerve lesion. *Neurosci. Lett.* **436:** 259–264.

40. Decosterd, I. & C.J. Woolf. 2000. Spared nerve injury: an animal model of persistent peripheral neuropathic pain. *Pain* **87:** 149–158.

41. Lovell, J.A. *et al.* 2000. Strain differences in neuropathic hyperalgesia. *Pharmacol. Biochem. Behav.* **65:** 141–144.

42. Yoon, Y.W. *et al.* 1999. Different strains and substrains of rats show different levels of neuropathic pain behaviors. *Exp. Brain Res.* **129:** 167–171.

43. del Rey, A. *et al.* 2011. Chronic neuropathic pain-like behavior correlates with IL-1β expression and disrupts cytokine interactions in the hippocampus. *Pain* **152:** 2827–2835.

44. Al-Amin, H. *et al.* 2011. Chronic dizocilpine or apomorphine and development of neuropathy in two animal models II: effects on brain cytokines and neurotrophins. *Exp. Neurol.* **228:** 30–40.

45. Gabay, E. *et al.* 2011. Chronic blockade of interleukin-1 (IL-1) prevents and attenuates neuropathic pain behavior and spontaneous ectopic neuronal activity following nerve injury. *Eur. J. Pain* **15:** 242–248.

46. Wolf, G. *et al.* 2003. Impairment of interleukin-1 (IL-1) signaling reduces basal pain sensitivity in mice: genetic, pharmacological and developmental aspects. *Pain* **104:** 471–480.

47. Wang, Z. *et al.* 2008. Interleukin-1 beta of Red nucleus involved in the development of allodynia in spared nerve injury rats. *Exp. Brain Res.* **188:** 379–384.

48. Milligan, E.D. *et al.* 2003. Spinal glia and proinflammatory cytokines mediate mirror-image neuropathic pain in rats. *J. Neurosci.* **23:** 1026–1040.

49. Besedovsky, H.O. & A. del Rey. 2011. Central and peripheral cytokines mediate immune-brain connectivity. *Neurochem. Res.* **36:** 1–6.

50. Bellinger, F.P., S. Madamba & G.R. Siggins. 1993. Interleukin 1 beta inhibits synaptic strength and long-term potentiation in the rat CA1 hippocampus. *Brain Res.* **628:** 227–234.

51. Katsuki, H. *et al.* 1990. Interleukin-1 beta inhibits long-term potentiation in the CA3 region of mouse hippocampal slices. *Eur. J. Pharmacol.* **181:** 323–326.

52. Yirmiya, R. & I. Goshen. 2011. Immune modulation of learning, memory, neural plasticity and neurogenesis. *Brain Behav. Immun.* **25:** 181–213.

53. Apkarian, A.V. *et al.* 2004. Chronic pain patients are impaired on an emotional decision-making task. *Pain* **108:** 129–136.

54. Hu, Y. *et al.* 2010. Amitriptyline rather than lornoxicam ameliorates neuropathic pain-induced deficits in abilities of spatial learning and memory. *Eur. J. Anaesthesiol.* **27:** 162–168.

55. Ji, G. *et al.* 2010. Cognitive impairment in pain through amygdala-driven prefrontal cortical deactivation. *J. Neurosci.* **30:** 5451–5464.

56. Norman, G.J. *et al.* 2010. Stress and IL-1beta contribute to the development of depressive-like behavior following peripheral nerve injury. *Mol. Psychiatry* **15:** 404–414.

Ann. N.Y. Acad. Sci. ISSN 0077-8923

ANNALS OF THE NEW YORK ACADEMY OF SCIENCES

Issue: *Neuroimmunomodulation in Health and Disease*

Experimental endotoxemia as a model to study neuroimmune mechanisms in human visceral pain

Sven Benson, Harald Engler, Manfred Schedlowski, and Sigrid Elsenbruch

Institute of Medical Psychology and Behavioral Immunobiology, University Hospital Essen, University of Duisburg-Essen, Essen, Germany

Address for correspondence: Sven Benson, Ph.D., Institute of Medical Psychology & Behavioral Immunobiology, University Hospital Essen, Hufelandstr. 55, 45122 Essen, Germany. sven.benson@uk-essen.de

The administration of bacterial endotoxin (i.e., lipopolysaccharide, LPS) constitutes a well-established experimental approach to study the effects of an acute and transient immune activation on physiological, behavioral, and emotional aspects of sickness behavior in animals and healthy humans. However, little is known about possible effects of experimental endotoxemia on pain in humans. This knowledge gap is particularly striking in the context of visceral pain in functional as well as chronic-inflammatory gastrointestinal disorders. Although inflammatory processes have been implicated in the pathophysiology of visceral pain, it remains incompletely understood how inflammatory mediators interact with bottom–up (i.e., increased afferent input) and top–down (i.e., altered central pain processing) mechanisms of visceral hyperalgesia. Considering the recent findings of visceral hyperalgesia after LPS application in humans, in this review, we propose that experimental endotoxemia with its complex peripheral and central effects constitutes an experimental model to study neuroimmune communication in human pain research. We summarize and attempt to integrate relevant animal and human studies concerning neuroimmune communication in visceral and somatic pain, discuss putative mechanisms, and conclude with future research directions.

Keywords: visceral pain; hyperalgesia; endotoxemia; lipopolysaccharide; cytokines; irritable bowel syndrome

Relevance and scope

Although pain or discomfort arising from the gastrointestinal tract constitutes an important protective mechanism allowing conscious awareness of potential injury or illness of visceral organs, chronic pain may lead to substantial morbidity.[1,2] Indeed, chronic abdominal pain is one of the most frequent reasons for medical consultation.[3] This is of high relevance for the functional gastrointestinal disorders (FGIDs), which are characterized by chronic or recurrent abdominal pain.[1,3] Irritable bowel syndrome (IBS), the most prevalent diagnosis among the FGIDs, is defined by recurrent abdominal pain and disturbed bowel habits in the absence of identifiable organic cause.[4] With prevalence rates of up to 20% in the United Kingdom and the United States,[5,6] IBS accounts for nearly one-third of all consultations with gastroenterologists,[7] which illustrates its major individual, social, and socioeconomic impact.[8]

The etiology and pathophysiology of chronic abdominal pain in IBS remain incompletely understood. IBS etiology most likely involves biological, psychological, and social factors that contribute to disturbances of the brain–gut axis and altered illness behavior.[9] Increased sensitivity to visceral stimuli (i.e., visceral hyperalgesia or hypersensitivity) together with an increased attention to signals arising from the gastrointestinal tract (i.e., visceral hypervigilance) is proposed to play an important role in the etiology and pathophysiology of the condition. Current concepts of the pathophysiology involve central and peripheral mechanisms mediated by the central nervous system, autonomic and neuroendocrine systems, and the enteric nervous system.[4,6,12] Recently, growing evidence supports that peripheral immune mechanisms and disturbed neuroimmune communication could contribute to the pathophysiology of visceral hyperalgesia and IBS.[9–12] As an extension of our previous review[9]

doi: 10.1111/j.1749-6632.2012.06622.x

Ann. N.Y. Acad. Sci. 1262 (2012) 108–117 © 2012 New York Academy of Sciences.

and original report,[61] the scope of this review is to attempt an integration of findings concerning neuroimmune communication in visceral pain and findings from both animal and human studies addressing the association between immune activation and pain. Basing our decision on the concept of sickness behavior, we propose the route of experimental endotoxemia as an approach suitable to address open questions regarding the relevance and mechanisms of altered neuroimmune communication in visceral pain. Of note, other areas of interest in the pathophysiology of visceral pain, such as psychosocial factors, autonomic and neuroendocrine pathways fall beyond the scope of this paper, but were the major focus of our recent review paper.[9]

Correlational findings in patients with IBS

Evidence in support of a putative role of inflammatory processes in chronic visceral pain is derived from postinfectious IBS.[12–15] Metaanalyses revealed a six- to sevenfold increased risk of developing IBS symptoms as a sequel of a bacterial gastroenteritis.[16,17] Even though the initial episode of infection had remitted in these patients, persistent visceral pain,[14] decreased rectal pain thresholds,[18–20] and persistent increases in the number and activity of mucosal inflammatory cells[14,19] were documented. Indicators of immune activation have also been documented in IBS patients without the onset of an acute infection, including alterations in mucosal cytokine production, mast cell activation, increased numbers of enterochromaffin cells and intraepithelial lymphocytes, low-grade infiltration of T cells in the myenteric plexus with neuronal degeneration, increased fecal concentrations of the antimicrobial peptide beta-defensin-2, and increased capsaicin receptor TRPV1-expressing sensory fibers in rectal biopsies, along with data suggesting a genetic predisposition favoring an increased production of tumor necrosis factor (TNF)-α (reviewed in Refs. 9 and 21). The putative role of gut microbiota in disturbances of the brain–gut axis in IBS patients is increasingly recognized, and dysbiosis may play a critical role in immune activation and low-grade inflammation in a proportion of patients.[22]

An increasing body of literature documents that inflammatory processes in patients with IBS are not limited to the gut. Local inflammation of the intestine may indeed result in systemic immune activation,[12,15,23,24] likely mediated by an increased permeability of the intestinal mucosa.[24] Studies in IBS patients without a history of an infection revealed low-grade activation of the innate immune system, including systemic increases in inflammatory markers such as TNF-α, interleukin (IL)-1β, IL-6, and IL-8.[9,21] Importantly, levels of proinflammatory cytokines, analyzed in the serum of IBS patients, have been found to be associated with IBS symptom score.[25] Lipopolysaccharide (LPS)-stimulated production of proinflammatory cytokines by peripheral blood mononuclear cells was significantly increased in IBS patients with diarrhea-associated pain.[26] Consistently, levels of circulating high-sensitivity C-reactive protein—even though in the normal laboratory range—were reportedly increased in IBS patients and correlated with disease severity.[27] Together, these clinical observational studies clearly support the notion that visceral pain in IBS patients is associated with both local and (low-grade) systemic immune activation, leading to the hypothesis that proinflammatory mediators may constitute relevant factors in the development and/or maintenance of visceral hyperalgesia.[28]

Sickness behavior: a concept integrating immune activation and pain

Infections, for example, with bacteria as observed in postinfectious IBS patients, lead to an activation of the innate immune system along with the production and release of proinflammatory cytokines such as TNF-α and IL-6. These inflammatory mediators not only control local and systemic immune responses to pathogens, but also exert effects on the central nervous system relevant to pain.[29–32] These behavioral changes, collectively termed "sickness behavior,"[33] include reduced exploration, increased anxiety, cognitive dysfunction, and social withdrawal in rodents,[29,30,34,35] as well as depressed mood and cognitive impairments in humans.[30,36,37] Importantly, pain facilitation or hyperalgesia is viewed as an integral part of sickness behavior, aiming at restricting activity.[38–40] This is corroborated by a number of studies involving somatic pain models that have clearly documented the proalgesic properties of proinflammatory cytokines (for reviews, see Refs. 40–42) by acting directly on nociceptors or indirectly by stimulating the release of agents like prostaglandins.[42] At the same time, cytokine antagonists reportedly reduce pain or hyperalgesia both in animal models of inflammation[40] and in patients with inflammatory conditions like inflammatory bowel disease[43] or rheumatoid arthritis.[44] Moreover, the link between pain and psychosocial stress, which is commonly reported by patients with chronic pain syndromes, has been attributed to stress-induced increases in proinflammatory cytokines.[44] However, whereas the associations among pain, inflammation, and sickness behavior have been one research focus in somatic pain, very few studies have thus far addressed these complex interactions in visceral pain.

Experimental endotoxemia as a model to study sickness behavior and pain

One experimental approach to study the effects of an immune activation on pain is the administration of bacterial endotoxin (i.e., lipopolysaccharide, LPS). LPS, the major component of the outer membrane of Gram-negative bacteria, is a prototypic pathogen-associated molecular pattern that stimulates the synthesis and release of proinflammatory cytokines by tissue macrophages and circulating monocytes.[41,45,46] Experimental administration of LPS in rodents and healthy human subjects demonstrably leads to a transient immune activation, which is characterized by elevated levels of circulating TNF-α, IL-1β, and IL-6.[37,39,46] The release of these inflammatory mediators has been causally linked to the behavioral and psychological effects of endotoxin-induced inflammation.[30,37,39,47]

Studies implementing experimental endotoxin administration in humans have mostly assessed immunological mechanisms of acute inflammation as well as effects on neuroendocrine parameters, sleep, and neuropsychological functions (for reviews, see Refs. 37, 46, and 47). Whereas high doses of endotoxin (i.e., 4 ng/kg body weight) were found to induce a flu-like syndrome characterized by fever, chills, headaches, nausea, general malaise, and increased heart rate and blood pressure,[46,48] application of lower endotoxin doses (i.e., 0.2–0.8 ng/kg body weight) led to significant increases in plasma cytokine levels, however, typically without the evocation of the previously mentioned flu-like symptoms.[48–53] On the other hand, low endotoxin doses reportedly induced mood changes including increased anxiety, depressed mood, and fatigue.[49,54,55] Whereas a dose of 0.8 ng/kg body weight endotoxin led to changes in memory performance,[49,56] no effects on neuropsychological functions were found

for lower doses.[50,53] While the behavioral and neuropsychological effects of LPS administration are relatively well studied, there is now emerging evidence of LPS-induced changes in pain. Experimental administration of bacterial endotoxin led to pain facilitation for noxious stimuli (for review, see Ref. 57), including thermal hyperalgesia[58,59] and mechanical allodynia,[60] which are mediated by the release of proinflammatory cytokines.[39,57]

Effects of LPS on human visceral pain

Effects of LPS on human visceral pain remained unclear until we recently conducted the first human study aiming to test the hypothesis that LPS-induced systemic inflammation increases visceral pain sensitivity.[61] We implemented a randomized crossover design; subjects received an intravenous injection of either 0.4 ng/kg body weight endotoxin or saline. To assess rectal sensory and pain thresholds, pressure-controlled distensions were delivered using a rectal balloon system, a procedure commonly used to quantify visceral pain sensitivity in healthy subjects and IBS patients.[62–65] In addition, predefined visceral stimuli of various intensities were rated on visual analog scales. All pain assessments were accomplished 2 hours postinjection, since IL-6 (previously found to increase somatic pain sensitivity[34,45]) as well as other proinflammatory mediators reportedly peak around this time point.[50,51,66] The results revealed the expected acute inflammatory response indicated by significant increases in plasma IL-6 and TNF-α levels together with a moderate increase in body temperature. More interestingly, LPS-induced immune activation also led to an increased sensitivity to rectal distensions as reflected by significantly decreased visceral sensory and pain thresholds when compared with saline control. Additionally, visceral distensions at low pressures (i.e., liminal stimuli) were rated as more unpleasant and inducing increased urge-to-defecate in the LPS condition. Correlation analysis revealed that pain thresholds were significantly associated with IL-6 levels within the LPS condition. Hence, this study lends first experimental support for the relevance of inflammatory processes in the pathophysiology of visceral hyperalgesia in healthy humans.[61]

The validity and clinical relevance of the experimental LPS model in the context of visceral pain are corroborated by observations in IBS patients, who display increased levels of systemic LPS associated with disease activity.[67,68] It is further supported by animal data using visceral pain stimuli. Coelho *et al.* documented in a series of studies in rats that intraperitoneal LPS injection lowered the visceral pain threshold via a mechanism involving mast cell degranulation and IL-1β and TNF-α release,[69] and that intracerebroventricular injections of these cytokines reproduced the LPS-induced pain responses.[70] Additionally, a study implementing excellular recordings from jejunal afferent nerves in rats demonstrated that intravenous LPS led to temporary increased responses to mechanical distensions and a delayed but maintained increase in spontaneous afferent discharge, suggesting that both spinal and vagal afferents may be modulated by inflammatory stimuli.[71]

If indeed LPS-induced immune activation leads to visceral hypersensitivity, as documented by our recent human study,[61] and proinflammatory agents have algesic properties,[40–42] then the question arises, which mechanisms underlie these effects? The next step is to give a brief—and given the complexity of the topic—incomplete overview of peripheral and central pathways of immune-to-brain communication, relevant to the endotoxin model, in the pathophysiology of visceral hypersensitivity (illustrated in Fig. 1).

Mechanisms: peripheral neuroimmune communication

The number of neurons contained in the gut is comparable to the number of neurons in the spinal cord. The complexity of the enteric nervous system owes to the multitude of tasks of the gastrointestinal tract, including digestion and absorption, as well as elimination and recognition of potential harmful agents.[28,72] Gut viscera are innervated by intrinsic and extrinsic spinal and vagal afferents.[2,72] Whereas intrinsic primary afferent neurons (IPANs) refer information required for control of digestion to the enteric nervous system,[73] extrinsic spinal and vagal nerves transduce information from the gut to the brain.[28,72] The latter contribute to the autonomic regulation of gut function and to the perception of pain.[28]

Extrinsic afferents have been classified on the basis of mechanical and electrophysiological response properties, including low- and high-threshold mechanosensitive vagal and spinal afferents.[72] While low-threshold receptors encode colorectal distensions over a wide range of innocuous and noxious stimuli, high-threshold receptors are mainly nociceptors.[1,28] Importantly, low- and high-threshold mechanoreceptors can sensitize during inflammation, suggesting that low-threshold afferents can contribute to pain and that visceral distensions at pressures normally not evoking pain may be encoded at intensities that cause discomfort.[28,72,74] This may occur through the release of inflammatory or proalgesic mediators produced by local cells such as mast cells.[11,69] Subsets of vagal nerve terminals end in close proximity of mucosal immune cells,[11,72] and vagal afferent nerves are sensitive to proinflammatory cytokines released from macrophages and to mast cell-derived mediators such as substance P, histamine, serotonin, and corticotrophin releasing factor.[28,69,72] Clinical evidence for this pathway was provided by Barbara *et al.*,[75] who demonstrated that a close proximity of activated mast cells to colonic nerve fibers was significantly correlated with severity of abdominal pain or discomfort, and that mucosal mast cell mediators from IBS patients had excitatory effects on rat nociceptive visceral sensory nerves.[76] Further, capsaicin receptor TRPV1–expressing sensory fibers in rectosigmoid biopsies were significantly increased in IBS patients and associated with an abdominal pain score.[77] Recent findings of a significant correlation between TRPV1 nerve fibers and abdominal pain in quiescent inflammatory bowel disease with IBS-like symptoms[78] support the putative role of TRPV1 in chronic visceral pain.[9] These mechanisms may lead to decreased threshold of activation and increased excitability of primary afferents, and thus to peripheral sensitization.[1,2]

Mechanisms: from the periphery to the CNS

Inflammatory signals from the gut are carried by the previously mentioned two categories of extrinsic nerves, that is, spinal and vagal afferents.[79,80] Signals from primary spinal afferents converge at the level of the dorsal horn of the spinal cord and account for the referral experienced with visceral pain.[1] Second-order neurons signal to the brain, transmitting conscious sensation of visceral sensory input via the spinothalamic tract, sensory nuclei of the thalamus to the somatosensory cortices SI and SII, the anterior cingulate cortex (ACC), and the insula.[1]

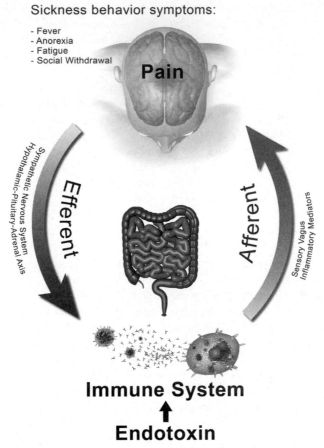

Sickness behavior symptoms:

- Fever
- Anorexia
- Fatigue
- Social Withdrawal

Pain

Hypothalamic-Pituitary-Adrenal Axis
Sympathetic Nervous System

Efferent

Afferent

Sensory Vagus
Inflammatory Mediators

Immune System

Endotoxin

Figure 1. Putative central and peripheral pathways that could mediate the effects of an acute peripheral immune activation, experimentally induced by the administration of bacterial endotoxin (i.e., lipopolysaccharide, LPS), on sickness behavior symptoms, including visceral pain and hyperalgesia.

The vagus is the most important visceral sensory nerve, and about 80–90% of the vagal axons are afferent nerve fibers.[28] Hence, the vagus nerve provides a major pathway for the transmission of information from the immune system to the brain.[81] Vagal nerve fibers terminate at the nucleus of the solitary tract (NTS), which projects to the thalamus and thereafter to cortical areas involved in visceral pain processing.[1,79] Additionally, vagal afferents project directly to areas involved in autonomic, emotional, and behavioral aspects of pain such as the hypothalamus, amygdala, locus coeruleus, and periaqueductal gray.[1,79,80] The pivotal role of vagal afferents in cytokine-induced sickness behavior, including impaired mood and pain facilitation, is well documented.[39,57,81] Indeed, intraperitoneal or intravenous application of bacterial endotoxin induces the activation of vagal afferents as indicated by the expression of c-Fos protein in va-

gal neurons.[82] Subdiaphragmatic vagotomy could block endotoxin-induced hyperalgesia in rats.[69] It has been proposed that, whereas spinal afferents primarily signal sensory aspects of visceral pain, vagus neurons may rather mediate autonomic, emotional-affective, and behavioral aspects of visceral nociception.[28,80,83] This is corroborated by a recent study from our laboratory analyzing the response of the amygdala to an endotoxin-induced systemic inflammation in rats.[29] Employing intracerebral electroencephalography and microdialysis in freely moving animals, this study revealed that intraperitoneal administration of LPS led to an increase in amygdaloid neuronal activity, and to *de novo* synthesis of proinflammatory cytokines in the amygdala. The changes in amygdala activation were related to an increase in anxiety-related behavior and reduced locomotor activity and exploration. These results provide evidence that the amygdala integrates immune-derived information to coordinate behavioral and autonomic responses.[29]

Other, nonneural mechanisms by which peripheral cytokines can reach the CNS may also be of relevance, since elevated levels of circulating proinflammatory cytokines have been observed in IBS patients and were associated with IBS symptoms, including visceral pain.[25,26] Peripheral proinflammatory cytokines can reportedly affect neural processing of ascending visceral input by directly reaching the brain via active transport, by diffusion from brain areas with incomplete blood–brain barrier, or by binding to cytokine receptors on endothelial cells of the brain microvasculature, which induce the production of secondary messengers, then diffuse to the brain and alter neural activity.[38,84]

Mechanisms: linking central neuroimmune mechanisms and pain

As mentioned previously, bacterial stimuli such as LPS and proinflammatory mediators may signal the brain via spinal, vagal, and humoral pathways, leading to characteristic behavioral and psychological changes collectively termed sickness behavior. One striking finding is that both animals and humans consistently show increased symptoms (or in the case of animals, indicators) of anxiety or depression in response to an experimental systemic immune activation. What are the neural circuits underlying these mood changes in sickness behavior? Animal studies employing bacterial stimuli to simulate gut infections showed that anxiety-like behavior was associated with activation of the NTS (i.e., the first central relay of vagal afferent fibers), the ventrolateral medulla, hypothalamus, central nucleus of the amygdala, and bed nucleus of the stria terminalis,[29,85] (for review, see Ref. 80). Interestingly, the latter three areas are involved in the integration of affective states such as anxiety or stress with autonomic and neuroendocrine responses.[29,80] Consistent with animal findings, human endotoxin studies could document mood changes including increased anxiety and depressiveness following endotoxin application,[49,54,55,66,86] mediated by proinflammatory cytokines.[30] Most recently, functional magnetic resonance imaging (fMRI) studies provided the first insight into the neural circuits involved in emotion processing during experimental endotoxemia. Eisenberger *et al.* reported a significant association between LPS-induced IL-6 increases and activity in the dorsal anterior cingulate cortex (dACC) and the anterior insula during a social exclusion task.[86] An fMRI study from our group designed to address neural responses to emotional stimuli during LPS-induced immune activation revealed an increased activation of prefrontal regions (i.e., inferior orbitofrontal, medial, and superior prefrontal cortices), which are closely connected to the amygdala and are involved in reappraisal and cognitive control of emotions.[66] Additionally, activation of the hypothalamus, supramarginal gyrus, parietal lobule, and middle occipital gyrus were observed.[66] Finally, Harrison *et al.* analyzed the effects of a typhoid vaccination on mood and neural mechanisms during an emotion task.[87] Of note, the typhoid model does not evoke increases in TNF-α and IL-1β, which play a key role in sickness behavior.[37] It is nevertheless of interest that mood impairments were associated with increased subgenual anterior cingulate cortex (sACC) activity and presumably with reduced connectivity of sACC to amygdala, medial prefrontal cortex, nucleus accumbens, and superior temporal sulcus in this study.[87]

Together, fMRI studies in human endotoxemia complement results from animal findings and help to elucidate the neural correlates of mood impairments in sickness behavior, including modified activation of the insula, ACC, and prefrontal regions. The pivotal role of these regions in emotional processing has been well established within the broad field of affective neurosciences.[88] Intriguingly, these brain areas are not only implicated in emotion processing, but are also known to process (visceral) sensory input,[1,79,89] thus providing a neurobiological link between pain and emotions with the CNS.[79,80,90] Studies in visceral and somatic pain indicate that in addition to primary pain processing networks, that is, the thalamus, insula, and somatosensory cortices SI and SII, activations in subregions of the cingulate and prefrontal cortices, amygdala, and hypothalamus mediate the various cognitive, affective, and motivational dimensions of the pain response.[1,89] Importantly, these pain-modulatory networks are highly relevant in functional gastrointestinal disorders. A recent quantitative meta-analysis of 18 studies comparing brain responses to rectal distension in patients with IBS and healthy controls revealed greater activations in amygdala and pregenual ACC (pACC) (i.e., in networks involved in emotional arousal and cortical pain modulation) and in anterior midcingulate cortex in IBS patients.[91] Notably, these differences cannot completely be explained by an increased afferent input from the gut, since similar differences have been documented during the anticipation of pain or sham distensions.[91] The pivotal role of emotions in visceral pain processing is further corroborated by a recent study of our group. We observed increased activation of the insula and prefrontal cortex during painful visceral stimulation in IBS patients compared with controls; however, these differences were abolished after controlling for differences in anxiety and depression scores between the groups.[64] Furthermore, within the group of IBS patients, anxiety scores were associated with pain-induced activation of the pACC and the aMCC.[64] The latter receives input from the amygdala and is thought to mediate the fear-avoidance aspect of pain. Together, these studies support that affective-emotional aspects of pain are crucial in explaining differences in visceral pain processing between IBS patients and healthy subjects.[64] Interestingly, the neural networks that have been implicated in the affective-motivational components of the response to pain in IBS patients[1,79,89] on the one hand, and altered emotion processing during experimental endotoxemia in healthy human subjects[66,86] on the other hand, appear to overlap. This poses the intriguing possibility that these neural networks may also mediate effects of inflammation on visceral hyperalgesia at the level of the CNS in chronic abdominal pain and IBS. Unfortunately, thus far, the putative association between affective-emotional pain processing and inflammation in IBS patients has not been addressed, but this is very clearly one intriguing area of future research.

Conclusions and future directions

The mechanisms mediating visceral pain may involve local, bottom–up as well as top–down processes, which, in concert, affect the balance between pain facilitation and inhibition (Fig. 1).[3,90] Inflammatory processes have been implicated in the pathophysiology of chronic visceral pain in conditions such as IBS, and may contribute via peripheral and CNS mechanisms to pain sensitization.[1] It remains incompletely clarified how inflammation affects the intricate relationship between bottom–up (i.e., afferent input) and top–down (i.e., central pain response) processes.[89] Experimental endo-

toxemia constitutes one interesting model to study neuroimmune communication in human visceral pain. Experimental application of endotoxin induces a transient low-grade systemic immune activation, mediated by the release of proinflammatory cytokines, and processed via vagal, spinal, and humoral pathways to the brain, resulting in sickness behavior.[46,47] Endotoxin-induced inflammation also reportedly lowered visceral sensory and pain thresholds in healthy humans,[61] corroborating the applicability of the model in visceral pain research. Moreover, neurophysiological and recent human brain imaging studies have proven the effects of experimental endotoxemia on neural networks implicated in emotion processing,[66,80,86] which intriguingly overlap with brain areas involved in visceral pain processing.[80] Future experimental endotoxemia studies implementing brain imaging techniques may not only help to further elucidate the relevance of inflammation for visceral pain, but also to help to disentangle the complex interactions of peripheral and central mechanisms, including sensory and affective-motivational networks, in the pathophysiology of visceral pain.

Conflicts of interest

The authors declare no conflicts of interest.

References

1. Knowles, C.H. & Q. Aziz. 2009. Basic and clinical aspects of gastrointestinal pain. *Pain* **141:** 191–209.
2. Akbar, A., J.R.F. Walters & S. Ghosh. 2009. Review article: visceral hypersensitivity in irritable bowel syndrome. Molecular mechanisms and therapeutic agents. *Alimentary Pharmacol. Therapeutics* **30:** 423–435.
3. Wilder-Smith, C.H. 2011. The balancing act: endogenous modulation of pain in functional gastrointestinal disorders. *Gut* **60:** 1589–1599.
4. Longstreth, G.F., W.G. Thompson, W.D. Chey, *et al.* 2006. Functional bowel disorders. *Gastroenterology* **130:** 1480–1491.
5. Jones, R. & S. Lydeard. 1992. Irritable-bowel-syndrome in the general-population. *Br. Med. J.* **304:** 87–90.
6. Talley, N.J., A.R. Zinsmeister, C. Vandyke & L.J. Melton. 1991. Epidemiology of colonic symptoms and the irritable-bowel-syndrome. *Gastroenterology* **101:** 927–934.
7. Mitchell, C.M. & D.A. Drossman. 1987. Survey of the aga membership relating to patients with functional gastrointestinal disorders. *Gastroenterology* **92:** 1282–1284.
8. Talley, N.J. 2008. Functional gastrointestinal disorders as a public health problem. *Neurogastroenterol. Motility* **20:** 121–129.
9. Elsenbruch, S. 2011. Abdominal pain in irritable bowel syndrome: a review of putative psychological, neural and neuroimmune mechanisms. *Brain Behav. Immun.* **25:** 386–394.

10. Bercik, P., E.F. Verdu & S.M. Collins. 2005. Is irritable bowel syndrome a low-grade inflammatory bowel disease? *Gastroenterol. Clin. N. Am.* **34:** 235–245.

11. O'Malley, D., E.M.M. Quigleya, T.G. Dinana & J.F. Cryana. 2011. Do interactions between stress and immune responses lead to symptom exacerbations in irritable bowel syndrome? *Brain Behav. Immun.* **25:** 1333–1341.

12. Spiller, R.C. 2009. Overlap between irritable bowel syndrome and inflammatory bowel disease. *Digestive Dis.* **27:** 48–54.

13. Collins, S.M., T. Piche & P. Rampal. 2001. The putative role of inflammation in the irritable bowel syndrome. *Gut* **49:** 743–745.

14. Spiller, R. & K. Garsed. 2009. Postinfectious irritable bowel syndrome. *Gastroenterology* **136:** 1979–1988.

15. Spiller, R.C., D. Jenkins, J.P. Thornley, *et al.* 2000. Increased rectal mucosal enteroendocrine cells, T lymphocytes, and increased gut permeability following acute campylobacter enteritis and in post-dysenteric irritable bowel syndrome. *Gut* **47:** 804–811.

16. Halvorson, H.A., C.D. Schlett & M.S. Riddle. 2006. Postinfectious irritable bowel syndrome—a meta-analysis. *Am. J. Gastroenterol.* **101:** 1894–1899.

17. Thabane, M., D.T. Kottachchi & J.K. Marshall. 2007. Systematic review and meta-analysis: the incidence and prognosis of post-infectious irritable bowel syndrome. *Alimentary Pharmacol. Therapeutics* **26:** 535–544.

18. Bergin, A.J., T.C. Donnelly, M.W. McKendrick & N.W. Read. 1993. Changes in anorectal function in persistent bowel disturbance following salmonella gastroenteritis. *Eur. J. Gastroenterol. Hepatol.* **5:** 617–620.

19. Gwee, K.A., Y.L. Leong, C. Graham, *et al.* 1999. The role of psychological and biological factors in postinfective gut dysfunction. *Gut* **44:** 400–406.

20. Mearin, F., A. Perello, A. Balboa, *et al.* 2009. Pathogenic mechanisms of postinfectious functional gastrointestinal disorders: results 3 years after gastroenteritis. *Scandinavian J. Gastroenterol.* **44:** 1173–1185.

21. Al-Khatib, K. & H.C. Lin. 2009. Immune activation and gut microbes in irritable bowel syndrome. *Gut Liver* **3:** 14–19.

22. Collins, S.M. & P. Bercik. 2009. The relationship between intestinal microbiota and the central nervous system in normal gastrointestinal function and disease. *Gastroenterology* **136:** 2003–2014.

23. Adam, B., T. Liebregts, J.M. Gschossmann, *et al.* 2006. Severity of mucosal inflammation as a predictor for alterations of visceral sensory function in a rat model. *Pain* **123:** 179–186.

24. Zhou, Q., B. Zhang & G.N. Verne. 2009. Intestinal membrane permeability and hypersensitivity in the irritable bowel syndrome. *Pain* **146:** 41–46.

25. Dinan, T.G., G. Clarke, E.M.M. Quigley, *et al.* 2008. Enhanced cholinergic-mediated increase in the proinflammatory cytokine IL-6 in irritable bowel syndrome: role of muscarinic receptors. *Am. J. Gastroenterol.* **103:** 2570–2576.

26. Liebregts, T., B. Adam, C. Bredack, *et al.* 2007. Immune activation in patients with irritable bowel syndrome. *Gastroenterology* **132:** 913–920.

27. Hod, K., R. Dickman, A. Sperber, *et al.* 2011. Assessment of high-sensitivity crp as a marker of micro-inflammation in irritable bowel syndrome. *Neurogastroenterol. Motility* **23:** 1105–1110.

28. Holzer, P. 2002. Sensory neurone responses to mucosal noxae in the upper gut: relevance to mucosal integrity and gastrointestinal pain. *Neurogastroenterol. Motility* **14:** 459–475.

29. Engler, H., R. Doenlen, A. Engler, *et al.* 2011. Acute amygdaloid response to systemic inflammation. *Brain Behav. Immun.* **25:** 1384–1392.

30. Dantzer, R., J.C. O'Connor, G.G. Freund, *et al.* 2008. From inflammation to sickness and depression: when the immune system subjugates the brain. *Nature Rev. Neurosci.* **9:** 46–57.

31. Raison, C.L., L. Capuron & A.H. Miller. 2006. Cytokines sing the blues: inflammation and the pathogenesis of depression. *Trends Immunol.* **27:** 24–31.

32. Sternberg, E.M. 2006. Neural regulation of innate immunity: a coordinated nonspecific host response to pathogens. *Nature Rev. Immunol.* **6:** 318–328.

33. Kent, S., R.M. Bluthe, K.W. Kelley & R. Dantzer. 1992. Sickness behavior as a new target for drug development. *Trends Pharmacol. Sci.* **13:** 24–28.

34. Cunningham, C., S. Campion, K. Lunnon, *et al.* 2009. Systemic inflammation induces acute behavioral and cognitive changes and accelerates neurodegenerative disease. *Biol. Psychiatr.* **65:** 304–312.

35. Sparkman, N.L., R.A. Kohman, A.K. Garcia & G.W. Boehm. 2005. Peripheral lipopolysaccharide administration impairs two-way active avoidance conditioning in c57bl/6j mice. *Physiol. Behav.* **85:** 278–288.

36. DellaGioia, N. & J. Hannestad. 2010. A critical review of human endotoxin administration as an experimental paradigm of depression. *Neurosci. Biobehav. Rev.* **34:** 130–143.

37. Kullmann, J.S., J.-S. Grigoleit & M. Schedlowski. 2011. Effects of an acute inflammation on memory performance, mood and brain activity. *Zeitschrift fuer Medizinische Psychologie* **20:** 108–117.

38. Maier, S.F. & L.R. Watkins. 2003. Immune-to-central nervous system communication and its role in modulating pain and cognition: implications for cancer and cancer treatment. *Brain Behav. Immun.* **17:** S125–S131.

39. Watkins, L.R. & S.F. Maier. 2005. Immune regulation of central nervous system functions: from sickness responses to pathological pain. *J. Intern. Med.* **257:** 139–155.

40. Ueceyler, N., M. Schaefers & C. Sommer. 2009. Mode of action of cytokines on nociceptive neurons. *Exp. Brain Res.* **196:** 67–78.

41. Miller, R.J., H. Jung, S.K. Bhangoo & F.A. White. 2009. Cytokine and chemokine regulation of sensory neuron function. *Handbook of Experimental Pharmacology*, 417–449. Springer.

42. Kidd, B.L. & L.A. Urban. 2001. Mechanisms of inflammatory pain. *Br. J. Anaesthesia* **87:** 3–11.

43. Ito, H. 2005. Treatment of crohn's disease with anti-IL-6 receptor antibody. *J. Gastroenterol.* **40:** 32–34.

44. Fitzcharles, M.A., A. Almahrezi & Y. Shir. 2005. Pain—understanding and challenges for the rheumatologist. *Arthritis Rheumatism* **52:** 3685–3692.

45. Akira, S. & K. Takeda. 2004. Toll-like receptor signalling. *Nature Rev. Immunol.* **4:** 499–511.

46. Bahador, M. & A.S. Cross. 2007. From therapy to experimental model: a hundred years of endotoxin administration to human subjects. *J. Endotoxin Res.* **13:** 251–279.

47. Yirmiya, R. & I. Goshen. 2011. Immune modulation of learning, memory, neural plasticity and neurogenesis. *Brain Behav. Immun.* **25:** 181–213.

48. Suffredini, A.F., H.D. Hochstein & F.G. McMahon. 1999. Dose-related inflammatory effects of intravenous endotoxin in humans: evaluation of a new clinical lot of Escherichia coli O:113 endotoxin. *J. Infect. Dis.* **179:** 1278–1282.

49. Reichenberg, A., R. Yirmiya, A. Schuld, *et al.* 2001. Cytokine-associated emotional and cognitive disturbances in humans. *Arch. Gen. Psychiatr.* **58:** 445–452.

50. Grigoleit, J.-S., J.R. Oberbeck, P. Lichte, *et al.* 2010. Lipopolysaccharide-induced experimental immune activation does not impair memory functions in humans. *Neurobiol. Learn. Mem.* **94:** 561–567.

51. Grigoleit, J.-S., J.S. Kullmann, O.T. Wolf, *et al.* 2011. Dose-dependent effects of endotoxin on neurobehavioral functions in humans. *PLoS ONE* **6:** e28330.

52. Krogh-Madsen, R., K. Moller, F. Dela, *et al.* 2004. Effect of hyperglycemia and hyperinsulinemia on the response of IL-6, TNF-alpha, and FFAs to low-dose endotoxemia in humans. *Am. J. Physiol. Endocrinol. Metabol.* **286:** E766–E772.

53. Krabbe, K.S., A. Reichenberg, R. Yirmiya, *et al.* 2005. Low-dose endotoxemia and human neuropsychological functions. *Brain Behav. Immun.* **19:** 453–460.

54. Eisenberger, N.I., E.T. Berkman, T.K. Inagaki, *et al.* 2010. Inflammation-induced anhedonia: endotoxin reduces ventral striatum responses to reward. *Biol. Psychiatr.* **68:** 748–754.

55. Eisenberger, N.I., T.K. Inagaki, N.M. Mashal & M.R. Irvin. 2010. Inflammation and social experience: an inflammatory challenge induces feelings of social disconnection in addition to depressed mood. *Brain Behav. Immun.* **24:** 558–563.

56. Cohen, O., A. Reichenberg, C. Perry, *et al.* 2003. Endotoxin-induced changes in human working and declarative memory associate with cleavage of plasma "readthrough" acetylcholinesterase. *J. Mol. Neurosci.* **21:** 199–212.

57. Watkins, L.R. & S.F. Maier. 2000. The pain of being sick: implications of immune-to-brain communication for understanding pain. *Ann. Rev. Psychol.* **51:** 29–57.

58. Mason, P. 1993. Lipopolysaccharide induces fever and decreases tail-flick latency in awake rats. *Neurosci. Lett.* **154:** 134–136.

59. Watkins, L.R., S.F. Maier & L.E. Goehler. 1995. Immune activation: the role of proinflammatory cytokines in inflammation, illness responses and pathological pain states. *Pain* **63:** 289–302.

60. Loram, L.C., F.R. Taylor, K.A. Strand, *et al.* 2011. Prior exposure to glucocorticoids potentiates lipopolysaccharide induced mechanical allodynia and spinal neuroinflammation. *Brain Behav. Immun.* **25:** 1408–1415.

61. Benson, S., J. Kattoor, A. Wegner, *et al.* 2012. Acute experimental endotoxemia induces visceral hypersensitivity and altered pain evaluation in healthy humans. *Pain* **153:** 382–390.

62. Benson, S., V. Kotsis, C. Rosenberger, *et al.* 2012. Behavioral and neural correlates of visceral pain sensitivity in healthy males and females: does sex matter? *Eur. J. Pain* **16:** 349–358.

63. Elsenbruch, S., C. Rosenberger, U. Bingel, *et al.* 2010. Patients with irritable bowel syndrome have altered emotional modulation of neural responses to visceral stimuli. *Gastroenterology* **139:** 1310–1318.

64. Elsenbruch, S., C. Rosenberger, P. Enck, *et al.* 2010. Affective disturbances modulate the neural processing of visceral pain stimuli in irritable bowel syndrome: an fMRI study. *Gut* **59:** 489–495.

65. Rosenberger, C., S. Elsenbruch, A. Scholle, *et al.* 2009. Effects of psychological stress on the cerebral processing of visceral stimuli in healthy women. *Neurogastroenterol. Motility* **21:** E740–E745.

66. Kullmann, J.S., J.-S. Grigoleit, P. Lichte, *et al.* 2012. Neural response to emotional stimuli during experimental human endotoxemia. *Hum. Brain Mapp.* Mar 28. doi: 10.1002/hbm.22063. [Epub ahead of print].

67. Gardiner, K.R., M.I. Halliday, G.R. Barclay, *et al.* 1995. Significance of systemic endotoxemia in inflammatory bowel disease. *Gut* **36:** 897–901.

68. Pastor Rojo, O., A. Lopez San Roman, E. Albeniz Arbizu, *et al.* 2007. Serum lipopolysaccharide-binding protein in endotoxemic patients with inflammatory bowel disease. *Inflamm. Bowel Dis.* **13:** 269–277.

69. Coelho, A.M., J. Fioramonti & L. Bueno. 2000. Systemic lipopolysaccharide influences rectal sensitivity in rats: role of mast cells, cytokines, and vagus nerve. *Am. J. Physiol.-Gastrointestinal Liver Physiol.* **279:** G781–G790.

70. Coelho, A.M., J. Fioramonti & L. Bueno. 2000. Brain interleukin-1 beta and tumor necrosis factor-alpha are involved in lipopolysaccharide-induced delayed rectal allodynia in awake rats. *Brain Res. Bull.* **52:** 223–228.

71. Liu, C.Y., W. Jiang, M.H. Muller, *et al.* 2005. Sensitization of mesenteric afferents to chemical and mechanical stimuli following systemic bacterial lipopolysaccharide. *Neurogastroenterol. Motility* **17:** 89–101.

72. Mayer, E.A. 2011. Gut feelings: the emerging biology of gut-brain communication. *Nature Rev. Neurosci.* **12:** 453–466.

73. Furness, J.B., W.A.A. Kunze, P.P. Bertrand, *et al.* 1998. Intrinsic primary afferent neurons of the intestine. *Prog. Neurobiol.* **54:** 1–18.

74. Gold, M.S. & G.F. Gebhart. 2010. Nociceptor sensitization in pain pathogenesis. *Nature Med.* **16:** 1248–1257.

75. Barbara, G., V. Stanghellini, R. De Giorgio, *et al.* 2004. Activated mast cells in proximity to colonic nerves correlate with abdominal pain in irritable bowel syndrome. *Gastroenterology* **126:** 693–702.

76. Barbara, G., B. Wang, V. Stanghellini, *et al.* 2007. Mast cell-dependent excitation of visceral-nociceptive sensory neurons in irritable bowel syndrome. *Gastroenterology* **132:** 26–37.

77. Akbar, A., Y. Yiangou, P. Facer *et al.* Increased capsaicin receptor TRPV1-expressing sensory fibres in irritable bowel syndrome and their correlation with abdominal pain. *Gut* **57:** 923–929.

78. Akbar, A., Y. Yiangou, P. Facer, *et al.* 2010. Expression of the trpv1 receptor differs in quiescent inflammatory bowel disease with or without abdominal pain. *Gut* **59:** 767–774.

79. Anand, P., Q. Aziz, R. Willert & L. Van Oudenhove. 2007. Peripheral and central mechanisms of visceral sensitization in man. *Neurogastroenterol. Motility* **19:** 29–46.

80. Goehler, L.E., M. Lyte & R.P.A. Gaykema. 2007. Infection-induced viscerosensory signals from the gut enhance anxiety: implications for psychoneuroimmunology. *Brain Behav. Immun.* **21:** 721–726.

81. Goehler, L.E., R.P.A. Gaykema, M.K. Hansen, *et al.* 2000. Vagal immune-to-brain communication: a visceral chemosensory pathway. *Autonomic Neurosci. Basic Clin.* **85:** 49–59.

82. Gaykema, R.P.A., L.E. Goehler, *et al.* 1998. Bacterial endotoxin induces fos immunoreactivity in primary afferent neurons of the vagus nerve. *Neuroimmunomodulation* **5:** 234–240.

83. Traub, R.J., J.N. Sengupta & G.F. Gebhart. 1996. Differential c-fos expression in the nucleus of the solitary tract and spinal cord following noxious gastric distention in the rat. *Neuroscience* **74:** 873–884.

84. Zhang, J.H. & Y.G. Huang. 2006. The immune system: a new look at pain. *Chinese Med. J.* **119:** 930–938.

85. Gaykema, R.P.A., L.E. Goehler & M. Lyte. 2004. Brain response to cecal infection with Campylobacter jejuni: analysis with Fos immunohistochemistry. *Brain Behav. Immun.* **18:** 238–245.

86. Eisenberger, N.I., T.K. Inagaki, L.T. Rameson, *et al.* 2009. An fmri study of cytokine-induced depressed mood and social pain: the role of sex differences. *Neuroimage* **47:** 881–890.

87. Harrison, N.A., L. Brydon, C. Walker, *et al.* 2009. Inflammation causes mood changes through alterations in subgenual cingulate activity and mesolimbic connectivity. *Biol. Psychiatr.* **66:** 407–414.

88. Phan, K.L., T.D. Wager, S.F. Taylor & I. Liberzon. 2009. Functional neuroimaging studies of human emotions. *CNS Spectrums* **9:** 258–266.

89. Azpiroz, F., M. Bouin, M. Camilleri, *et al.* 2007. Mechanisms of hypersensitivity in ibs and functional disorders. *Neurogastroenterol. Motility* **19:** 62–88.

90. Wiech, K. & I. Tracey. 2009. The influence of negative emotions on pain: behavioral effects and neural mechanisms. *NeuroImage* **47:** 987–994.

91. Tillisch, K., E.A. Mayer & J.S. Labus. 2011. Quantitative meta-analysis identifies brain regions activated during rectal distension in irritable bowel syndrome. *Gastroenterology* **140:** 91–100.

Ann. N.Y. Acad. Sci. ISSN 0077-8923

ANNALS OF THE NEW YORK ACADEMY OF SCIENCES
Issue: *Neuroimmunomodulation in Health and Disease*

The neuroimmune connection interferes with tissue regeneration and chronic inflammatory disease in the skin

Eva M.J. Peters,[1,2] Christiane Liezmann,[2] Burghard F. Klapp,[1] and Johannes Kruse[2]

[1]Center for Internal Medicine and Dermatology, Charité-University Medicine, Berlin, Germany. [2]Department of Psychosomatic Medicine, Justus-Liebig University, Giessen, Germany

Address for correspondence: Eva. M.J. Peters, Laboratory for Psychoneuroimmunology, Ludwigstr. 78, D-35392, Giessen, Germany. eva.peters@eva-peters.com

Research over the past decades has revealed close interactions between the nervous and immune systems that regulate peripheral inflammation and link psychosocial stress with chronic somatic disease. Besides activation of the sympathetic and the hypothalamus–pituitary–adrenal axis, stress leads to increased neurotrophin and neuropeptide production in organs at the self–environment interface. The scope of this short review is to discuss key functions of these stress mediators in the skin, an exemplary stress-targeted and stress-sensitive organ. We will focus on the skin's response to acute and chronic stress in tissue regeneration and pathogenesis of allergic inflammation, psoriasis, and skin cancer to illustrate the impact of local stress-induced neuroimmune interaction on chronic inflammation.

Keywords: stress; neuroimmunology; psychodermatology; neurotrophin/neuropeptide stress axis; neuroimmune plasticity

Introduction

Neuroimmune interactions link stress and the immune response during different physiologic and pathologic processes in organs at the border between our body and the environment, such as the skin. We shall briefly mention the well-known activation of the classical stress neuroendocrine responses such as those mediated by the sympathetic axis (SA) and the hypothalamus–pituitary–adrenal (HPA) axis. Our main scope, however, is to discuss the role of key neurotrophins and neuropeptides in response to various stressors, ranging from acute to chronic, and the relevance of these mediators for inflammatory processes. The role of neurotrophin and neuropeptide actions (NNA) in inflammation in the skin will be emphasized. For this purpose, physiologically occurring tissue remodeling processes, hair growth, allergic inflammation, and psoriasis (as two contrasting chronic pathologies), and skin cancer will be discussed as instructive examples.

Stress and immunological concepts

Effects of stress mediators have been commonly discussed with reference to the favored immunological concepts of the time.[1] The T helper 1/T helper 2 (Th1/Th2) hypothesis, for example, was generated on the basis of cell culture experiments, with the aim of explaining delayed type hypersensitivity as an example of cell-mediated tissue destruction as opposed to tissue protective B cell activation and antibody production.[1,2] The corresponding proinflammatory and cellular immune responses on the one hand, and anti-inflammatory and humoral immune responses on the other, were largely employed as synonyms and results of stress research were reported under their label.[3] Lately, however, the hypothesis of a Th1/Th2 misbalance aggravating inflammatory processes has been modified to accommodate Th17 cell and regulatory cell functions, and pro- and anti-inflammatory cytokines were shown to occasionally possess the opposing effect, depending on the model used.[1,4,5] Thus, it is more realistic to discuss cytokine effects with reference to their *in vivo* function under a defined condition. Also, inflammation status and background disease pathology are important issues and require stepwise analysis of neuroimmune involvement in the pathogenesis of disturbed tissue regeneration, chronic inflammation, and cancer.

doi: 10.1111/j.1749-6632.2012.06647.x

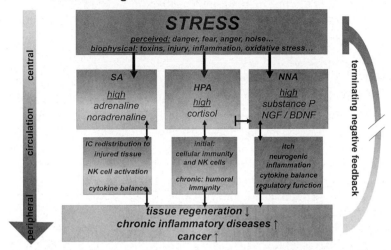

Figure 1. Schematic representation of immunological effects mediated by neuroendocrine responses to stress and their potential disruption during disease. The schematic drawing depicts selected key effects of pathogenically relevant stress effects acting through deregulated respective stress mediators on the immune system and subsequently on chronic disease. Under physiological conditions, a balanced release of stress mediators mediate adaptation to the stressor and the response is timely terminated. In the depicted pathological release, no adaptation to the stressor occurs, and the response is prolonged as indicated by the interrupted gray arrow. The drawing does not include the many pathophysiological effects of these mediators, e.g., on energy metabolism or sensation. HPA, hypothalamus pituitary adrenal axis; IC, cells of the immune system; NK, natural killer cells; NNA, neurotrophin neuropeptide axis; PA, parasympathetic axis; SA, sympathetic axis; Th, T helper cells.

The sympathetic stress axis alters peripheral immune function

Since Walter Cannon described its role in the "fight or flight response," the first acknowledged stress axis is the sympathetic axis (SA; Fig. 1),[6] though the term stress in its present use was coined later by Hans Selye. With time, it became evident that activation of this axis has immunological implications: upon acute stress, activation of the SA results, for example, in a rapid redistribution of the cells of the immune system into target organs;[7,8] and depending on the concentration of the mediator and the presence of specific receptors on target cells of the immune system, it is involved in modulation of innate as well as specific immune responses after stress exposure.[9] One of the most prominent effects is the migration of NK cells and the enhancement of neutrophil function and cellular immunity.[10,11] The receptors involved appear to be alpha-adrenoreceptors, since their activation promotes cytokine production such as TNF-α and IL-1β together with lymphocyte homing and neutrophilia.[12] In contrast, IL-10 and transforming growth factor beta (TGF-β) production as well as lymphocyte pro-

liferation are induced by beta-adrenoreceptor activation, which can override the activation of alpha receptors.[12] Moreover, receptor availability diverts between acute and chronic proinflammatory and anti-inflammatory effects of SA activation.

The hypothalamus–pituitary–adrenal stress axis and its role in keeping the immune balance

In the 1950s, a second stress axis became known, and the term *stress* in its present meaning was coined to describe the function of the HPA axis.[13] Today, a large number of immunological effects of the neuropeptide and endocrine mediators released after activation of this axis are known (Fig. 1). Depending on the secretion mode and amount, these neuroendocrine mediators act systemically as well as locally to shape an immune response that is optimally suited to adapt to either acute or chronic challenges.[4,5,14,15] Acute stress, for example, suppresses TNF-α, IFN-γ, or IL-12 due to the acute secretion of high amounts of glucocorticoids.[16,17] Chronic stress, by contrast, increases basal cortisol secretion together with a reduced acute excitability of the HPA axis, which appears to promote the production of IL-4, IL-10, and IL-13, and has therefore been discussed as a prime mechanism in stress-induced induction and aggravation of allergic and autoimmune disease.[18]

Neurotrophins and neuropeptides and their potential role in the regulation of stress and immune responses

Since the discovery of neurotrophins and neuropeptides, more and more data accumulate that document a prominent role of these mediators in stress and the existence of another stress axis may be postulated. This axis depends on neurotrophin signaling, employs peptidergic nerve fibers, and is always coactivated with the SA and HPA axis.[18,19] Through mediators such as substance P (SP), localized neuroimmune interactions can take place anywhere where nerve fibers meet with cells of the immune system, but it mainly occurs in peripheral organs at the self-environment interface or in immune-competent tissues.[18,20,21] Following stress, the resulting immunological changes are complex and include proinflammatory as well as anti-inflammatory effects.[18,22]

Corresponding to their initial discovery in peripheral sensory nerve fibers, one of the first stress-related functions described for neuropeptide mediators was the activation of neurogenic inflammation, which includes the release of neuropeptides and subsequent degranulation of mast cells, followed by enhanced endothelial permeability, plasma extravasation, and infiltration of the affected tissue. This innate neuroimmune response aggravates virtually all so far studied chronic inflammatory diseases of organs at the self-environment interface, irrespective of categories such as Th1 or Th2.[23–25]

In addition, neurotrophins and neuropeptides facilitate the transition to other innate as well as specific immune responses.[1] Besides its role in neurogenic inflammation, SP enhances NK cell activity and LFA-1 expression on CD8[+] cells,[10] promotes lymphocyte proliferation of CD4[+] T cells but not B cells,[26] and induces IFN-γ and TNF-α production.[27] Its production is induced in lymphocytes and macrophages by IL-12 and IL-23 and blocked by IL-10 and TGF-β.[28] Thus, SP supports acute inflammatory cellular immune responses. At the same time, SP induces the expression of CD4 together with CD25 and reduces IgE production.[26,29,30] Thus,

SP is also involved in the induction of T regulatory effects.[18]

Stress-induced release of the neurotrophin nerve growth factor (NGF) in addition to its local production in peripheral tissues in response to perceived as well as mechanical stress exposure was also shown.[31,32] On the one hand, NGF supports outgrowth of peripheral peptidergic nerve fibers, and thus neuronal plasticity, and it supports mast cell homeostasis. This further enhances the susceptibility to external and internal triggers of neurogenic inflammation such as noise or IgE.[18,33] On the other hand, NGF production is promoted by TNF-α and IL-1, and when increased, NGF activates T cells in inflamed tissue to produce cytokines such as IL-4.[34,35] NGF also promotes IL-10 production by monocytes,[36] enhances IL-4 production by eosinophils via the low-affinity pan-neurotrophin receptor p75NTR,[37,38] and can directly affect proliferation of keratinocytes, which express the high-affinity NGF receptor TrkA.[33] It is thus a key candidate for aggravated humoral, allergic, and hyperproliferative cutaneous inflammation after stress exposure.[38]

Brain-derived neurotrophic factor (BDNF) is another neurotrophin with a growing role in stress biology as well as in neuroimmune regulation. It is also produced in peripheral tissues, such as the skin,[39] but in contrast to NGF, it cannot exert direct growth-promoting effects on keratinocytes or naive T cells, as these express a nonfunctional isoform of the high-affinity receptor TrkB.[40] However, the stress response–modifying capacities of BDNF render it a highly potential player in both perceived stress and cutaneous inflammation. Acute stress increases BDNF levels, which corresponds to acute inflammatory states especially in allergic inflammation.[41,42] Moreover, BDNF is produced by monocytes in response to IL-6 and TNF-α,[43] and functional TrkB is expressed by stimulated T cells producing cytokines such as IFN-γ.[44] Thus, BDNF may be a good marker for acute stress and cellular inflammation.

The skin as a model of pathogenic neuroimmune interactions during stress

Stress, defined as the neuroendocrine response of the organism to any threat it encounters, which requires an adaption of the organisms' biopsychosocial functions in order to cope with it, was linked to the development and control of inflammatory diseases of the skin since the late 1960s and early 1970s of the past century. The initial experiments demonstrated a higher susceptibility to certain mucosal viral infections first in stressed mice and then in stressed humans,[45] and suggested increased susceptibility of organs at the self-environment interface to succumb to new inflammatory challenges. Commonly, this is attributed to altered HPA axis and SA function. However, recent data suggests a prominent role of locally acting neurotrophins and neuropeptides.

The skin is equipped with an abundant supply of nerve fibers that contact virtually all functional and structural cell populations found in the skin. Nerve fibers in the skin travel close to keratinocytes, fibroblasts, endothelial cells, Schwann cells, resident immune cells such as mast cells, and immune cells that invade peripheral tissue such as antigen-presenting dendritic cells or lymphocytes. These cell populations produce neurotrophins, and plasticity of peripheral innervation as well as of contacts between nerve fibers and cells of the immune system has been reported a number of times:[18] cutaneous neurotrophin levels as well as the number of detectable cutaneous neuropeptidergic nerve fibers and their contacts with cells of the immune system usually increase during stress as well as in chronic inflammatory states.[18] Moreover, immune cells are capable of producing not only neurotrophins but also some neuropeptides that can act in a paracrine fashion.[28,46] Altered secretion of neuotrophins and the described plasticity of neuroimmune interaction in the skin primarily facilitates neurogenic inflammation and promotes cutaneous inflammation in healthy as well as in diseased skin.[18]

Furthermore, peptidergic nerve fibers interact with antigen-presenting dendritic cells in the skin.[47] This interaction is triggered by acute challenge, either due to perceived stress or antigen, and initially recruits lymphocytes to the inflammation site.[48–50] However, following the initial activation and especially after repeated activation, neuropeptides such as SP frequently suppress antigen presentation and promote T regulatory cell (T_{reg} cell) function, thereby attenuating cutaneous inflammation.[30,47]

Peripheral tissue talks back to the brain
Neurotrophins, neuropeptides, and cytokines derived from the skin may also have an effect on the activation of the central stress axes. SP blocks the activation of the HPA axis at the level of the hypothalamus,[51] and SP and NA interact in a stress-dependent manner in the locus coeruleus and the medial prefrontal cortex—activation that can be blocked by peripheral SP-antagonist application.[52,53] Also, BDNF modifies the activity of the HPA axis,[54] and chronic stress and depression reduce the expression of BDNF in the hippocampus,[54] while repeated stress activation alternating with relaxation enhance its levels and modify peripheral SA activity.[55] These observations indicate that the activation of the central stress axis is intimately linked with peripheral immune activation in both directions, and that the NNA can regulate HPA axis as well as SA activation.

Hair growth: an instructive model for hampered tissue regeneration under stress
Hair growth is one of the few tissue regeneration processes that regularly takes place in a healthy, unperturbed mammalian organism. Tight interaction between the mesenchymal compartment, represented by the dermal papilla, and a multilayered epithelial compartment, which produces the hair shaft, regulates phases of growth (anagen), regression (catagen), and relative quiescence (telogen). Acting through peptidergic nerve fibers that accompany each hair follicle, neurotrophins, SP, and mast cell activation block healthy hair growth in mice with and without a predisposition to develop alopecia. This interaction is stress dependent.[56,57] The effect appears to depend on the generation of a proinflammatory state, including enhanced expression of TNF-α and activation of CD8$^+$ T cells. Stress-induced neurotrophins and neuropeptides also affect epithelial cells expressing the respective receptor combination directly (e.g., NK1 and p75NTR). Organ culture experiments employing human scalp skin hair follicles allow us for the conclusion that neurotrophins and neuropeptides also act in humans[58,59] and suggest that acute stress via neurogenic inflammation and cellular immunity hampers hair growth.

Atopic dermatitis: a chronic inflammatory disease driven by neurogenic and humoral immunity in response to stress
At a systemic level, numerous works by Buske-Kirschbaum *et al.* have impressively shown that healthy individuals with chronic stress and those who have an atopic diatheses share a decreased

reactivity of the HPA axis to acute stress exposure (c.f., Refs. 61, 62). This neuroendocrine state is associated with the production of cytokines such as IL-4 and IL-5, which are characteristic for atopic lesions and promote allergic humoral and eosinophilic inflammation.[60–62] An altered HPA function was thus thought to contribute to the pathogenesis and worsening of atopic inflammatory diseases upon stress exposure.[61–63]

In addition, pruritus, a key symptom of atopic dermatitis, not only greatly reduces the quality of life of the affected patients, with psychosocial, medical, and economic consequences,[18] but it indicates the release of neuropeptides, which produce itching and scratching behavior within minutes.[64] A vicious circle is set in motion that aggravates skin lesions and is linked to stress and malfunctioning stress adaptation, especially after excessive, uncontrollable, and unpredictable biophysical, as well as psychosocial, stress exposure.[18] From the mouse model, we can conclude that stress leads to the aforementioned neurotrophin-dependent neuronal plasticity and, hence, increased contacts between peptidergic nerve fibers and mast cells in allergic inflammation.[18,20,24] Interestingly, IL-4 further enhances mast cell responsiveness to SP,[65] and IgE-induced mast cell degranulation is facilitated in the presence of mast cell innervation.[66] In addition, IL-5 and vascular cell adhesion molecule (VCAM) are produced in a stress- and SP-dependent manner, which contributes to eosinophilia.[67] These changes alter the susceptibility of atopic/allergic inflammatory diseases to worsen upon stress exposure and to improve after pharmacological and psychosocial interventions that reduce the stress response.

Approximately 48 h after induction of inflammation, the cytokine profile in allergic inflammation changes and additional IFN-γ comes into play.[68] This increase is accompanied by IL-2 production, and can be explained by the interaction of peptidergic nerve fibers with dendritic cells in the maturing lesion.[30] Correspondingly, repeated stress exposure leads to IFN-γ production, SP-dependent T_{reg} cell induction, and subsequent suppression of allergic inflammation in a mouse model employing repeated stress exposure.[30] Thus, acute stress enhances allergic inflammation mainly through neurogenic inflammation, while in maturing lesions as well as under repeated stress exposure, neuroimmune plasticity is associated with a conditioning process aiming at termination of inflammation. In atopic disease and allergy, one could say, "A stressor a day keeps the doctor away." This observation holds implications for therapeutic options provided by the analysis and manipulation of neurotrophins and neuropeptides, for example, for the training aspect of stress adaptation.[69]

Psoriasis: stress-induced enhancement of cellular immunity

A correlation between stress and the onset and severity of psoriasis has been shown in retrospective studies.[70,71] Interestingly, compared with atopic/allergic disease, HPA axis malfunction appears to play a minor role in the sensitivity of psoriasis to stress-induced aggravation.[72] The number of publications that state altered HPA function in patients with psoriasis equals the number of publications that prove the opposite. In contrast, increased release of noradrenaline was repeatedly observed in stressed psoriatic patients.[73,74] This would fit with the cytokine production that predominates in psoriasis, characterized by high TNF-α and IFN-γ levels, as well as with enhanced lymphocyte migration.[75] TNF-α inhibitors indeed greatly affect the course of the disease.[76] In contrast, blockade of IFN-γ was not successful in the clinical setting and has led to a more stage-wise dynamic perspective, including, for example, Th17 immune responses.[75]

The neuroimmune relationship between psychological stress and psoriasis again hints at neurogenic inflammation (Fig. 1). Psoriatic complaints include itching, though not as prominently as in atopic dermatitis.[77] Lesional psoriatic skin contains significantly more nerve fibers that contain SP and SP+ nerve fiber contacts with NK1+ mast cells,[78,79] which can contribute to TNF-α release, than healthy skin.[80,81] Interestingly, low cortisol levels correlate with increased numbers of SP+ inflammatory cells in psoriatic lesions, indicating interaction between the local equivalent of the HPA with SP effects.[72] Altered expression of neurotrophins in chronically inflamed psoriatic skin has also been reported and expression of the proliferation-promoting NGF receptor TrkA spreads beyond the basal layer, while expression of the potentially apoptosis-inducing pan-neurotrophin receptor p75NTR is reduced.[82] Accordingly, keratinocyte hyperproliferation partially depends on NGF and, in addition, NGF activates T cells during psoriatic inflammation.[35] Thus, acute stress plays an aggravating role in this inflammatory disease driven by TNF-α and epithelial hyperplasia through SA and NNA promotion of acute inflammation.

Neuroimmune regulation of skin cancer: natural killer cell activation and blockers of T_{reg} cells are protective

Patients diagnosed with skin cancer, especially with melanoma, are under great pressure and experience considerable stress.[83] High stress levels in melanoma patients were associated with a favorable effect of a psychotherapeutic intervention on survival and NK cell function in stressed American[84] but not in relaxed Danish patients.[85,86] Apparently, stress-induced hyperactivation of the SA plays a cancer-promoting role[87] (Fig. 1), at least in a mouse model of melanoma, since propranolol prevents tumor growth and metastasis, for example, after surgery or perceived stress.[88,89] Similar effects can be expected in humans as melanoma cells express beta-adrenoreceptors.[90] Chronic stress may also be involved in skin cancer development through attenuation of the function of the HPA axis. As discussed previously, chronic activation of the HPA leads to decreased peak and increased basal cortisol production. This is associated with suppressed NK and cellular immune function and thus decreased capacity to eliminate cancer cells.[91–93]

Interestingly, seemingly paradoxical results were obtained in a mouse model both for epithelial skin tumors as well as for melanoma.[5] In contrast to chronic stress, acute stress increased IFN-γ and IL-12 and provided some resistance to squamous cell carcinoma development.[94] Such an effect may involve the NNA. Experimental mice are usually kept under rather unchallenging conditions. Mice kept in larger cages and provided with the opportunity to run, hide, and build (so-called enriched housing conditions) are protected from melanoma metastasis via the release of leptin from adipose tissue following BDNF-induced SA activation.[55] Thus, acute and repeated mild stress exposure through SA and NNA activation may serve a protective function, while chronic stress, through HPA modification, may enhance skin cancers.

Conclusions

We have discussed that by interacting and regulating each other, the SA, the HPA axis, and NNA conduct a many-voiced neuroimmune concert, which controls inflammation in response to diverse challenges in organs that, as the skin, are at the self-environment border. The effect of stress on skin disease, however, has mostly been investigated at the level of central stress responses along the HPA axis and SA.[3,46] This leads to rapid redistribution of immune cells to peripheral organs, activation of NK cells, and other cellular immune responses together with the modulation of the production of cytokines such as TNF-α. By contrast, an altered control of the reactivity of the HPA axis and SA in response to chronic stress may facilitate allergic disease and tumor development.[61,62] Stress also leads to deterioration of neurogenic inflammation, hampering tissue regeneration and promoting acute flare up of chronic inflammatory diseases, such as atopic dermatitis or psoriasis. This seems to depend on the communication between neurogenic and immunogenic components.[72,95] Repeated activation of neuropeptide and neurotrophin production, however, leads to T_{reg} cell induced attenuation of certain inflammatory processes. Specific stress-induced neuroimmune interactions that enhance distinct disease pathologies, and thus aggravate disease, can be influenced by pharmacological as well as by psychosocial interference, and improved adaptation to stress can be achieved with benefit for tissue regeneration, inflammatory disease, and cancer.

Conflicts of interest

The authors declare no conflicts of interest.

References

1. Steinman, L. 2007. A brief history of T(H)17, the first major revision in the T(H)1/T(H)2 hypothesis of T cell-mediated tissue damage. *Nat. Med.* **13:** 139–145.
2. Coffman, R.L. 2006. Origins of the T(H)1-T(H)2 model: a personal perspective. *Nat. Immunol.* **7:** 539–541.
3. Elenkov, I.J. 2008. Neurohormonal–cytokine interactions: implications for inflammation, common human diseases and well-being. *Neurochem. Int.* **52:** 40–51.
4. Rohleder, N. 2012. Acute and chronic stress induced changes in sensitivity of peripheral inflammatory pathways to the signals of multiple stress systems—2011 Curt Richter Award Winner. *Psychoneuroendocrinology* **37:** 307–316.
5. Dhabhar, F.S. 2009. Enhancing versus suppressive effects of stress on immune function: implications for immunoprotection and immunopathology. *Neuroimmunomodulation* **16:** 300–317.
6. Cannon, W.B. 1929. *Bodily Changes in Pain, Hunger, Fear, and Rage.* Appleton-Century-Crofts. New York.

7. Rogausch, H., A. del Rey, J. Oertel & H.O. Besedovsky. 1999. Norepinephrine stimulates lymphoid cell mobilization from the perfused rat spleen via beta-adrenergic receptors. *Am. J. Physiol.* **276:** R724–730.

8. Elenkov, I.J., R.L. Wilder, G.P. Chrousos & E.S. Vizi. 2000. The sympathetic nerve—an integrative interface between two supersystems: the brain and the immune system. *Pharmacol. Rev.* **52:** 595–638.

9. Ziemssen, T. & S. Kern. 2007. Psychoneuroimmunology—cross-talk between the immune and nervous systems. *J. Neurol.* **254:** II/8–II/11.

10. Lang, K., T.L. Drell, B. Niggemann, *et al.* 2003. Neurotransmitters regulate the migration and cytotoxicity in natural killer cells. *Imm. Lett.* **90:** 165–172.

11. Engler, H. *et al.* 2010. Chemical destruction of brain noradrenergic neurons affects splenic cytokine production. *J. Neuroimmunol.* **219:** 75–80.

12. Bergmann, M. & T. Sautner. 2002. Immunomodulatory effects of vasoactive catecholamines. *Wien. Klin. Wochenschr.* **114:** 752–761.

13. Seyle, H. 1950. *The Physiology and Pathology of Exposure to STRESS*. ACTA. Inc. Medical Publishers. Montreal.

14. Elenkov, I.J. & G.P. Chrousos. 2006. Stress system—organization, physiology and immunoregulation. *Neuroimmunomodulation* **13:** 257–267.

15. Irwin, M.R. 2008. Human psychoneuroimmunology: 20 years of discovery. *Brain. Behav. Immun.* **22:** 129–139.

16. Wang, Y. *et al.* 2008. Enhanced resistance of restraint-stressed mice to sepsis. *J. Immunol.* **181:** 3441–3448.

17. Webster, E.L., D.J. Torpy, I.J. Elenkov & G.P. Chrousos. 1998. Corticotropin-releasing hormone and inflammation. *Ann. N.Y. Acad. Sci.* **840:** 21–32.

18. Liezmann, C., B. Klapp & E.M. Peters. 2011. Stress, atopy and allergy: a re-evaluation from a psychoneuroimmunologic persepective. *Dermatoendocrinology* **3:** 37–40.

19. Rosenkranz, M.A. 2007. Substance P at the nexus of mind and body in chronic inflammation and affective disorders. *Psych. Bull.* **133:** 1007–1037.

20. Steinhoff, M. *et al.* 2003. Modern aspects of cutaneous neurogenic inflammation. *Arch. Dermatol.* **139:** 1479–1488.

21. Mignini, F., V. Streccioni & F. Amenta. 2003. Autonomic innervation of immune organs and neuroimmune modulation. *Auton. Autacoid. Pharmacol.* **23:** 1–25.

22. Elenkov, I.J. & G.P. Chrousos. 2002. Stress hormones, proinflammatory and antiinflammatory cytokines, and autoimmunity. *Ann. N.Y. Acad. Sci.* **966:** 290–303.

23. Singh, L.K., X. Pang, N. Alexacos, *et al.* 1999. Acute immobilization stress triggers skin mast cell degranulation via corticotropin releasing hormone, neurotensin, and substance P: a link to neurogenic skin disorders. *Brain. Behav. Immun.* **13:** 225–239.

24. Black, P.H. 2002. Stress and the inflammatory response: a review of neurogenic inflammation. *Brain. Behav. Immun.* **16:** 622–653.

25. Harvima, I.T., G. Nilsson & A. Naukkarinen. 2010. Role of mast cells and sensory nerves in skin inflammation. *G. Ital. Dermatol. Venereol.* **145:** 195–204.

26. Santoni, G., M.C. Perfumi, E. Spreghini, *et al.* 1999. Neurokinin type-1 receptor antagonist inhibits enhancement of T cell functions by substance P in normal and neuro-manipulated capsaicin-treated rats. *J. Neuroimmunol.* **93:** 15–25.

27. Levite, M. 2000. Nerve-driven immunity. The direct effects of neurotransmitters on T cell function. *Ann. N.Y. Acad. Sci.* **917:** 307–321.

28. Blum, A., T. Setiawan, L. Hang, *et al.* 2008. Interleukin-12 (IL-12) and IL-23 induction of substance p synthesis in murine T cells and macrophages is subject to IL-10 and transforming growth factor beta regulation. *Inf. Immunol.* **76:** 3651–3656.

29. Carucci, J.A., D.L. Auci, C.A. Herrick & H.G. Durkin. 1995. Neuropeptide-mediated regulation of hapten-specific IgE responses in mice: I. Substance P-mediated isotype-specific suppression of BPO-specific IgE antibody-forming cell responses induced in vivo and in vitro. *J. Leukoc. Biol.* **57:** 110–115.

30. Pavlovic, S. *et al.* 2011. Substance P is a key mediator of stress-induced protection from allergic sensitization via modified antigen presentation. *J. Immunol.* **186:** 848–855.

31. Peters, E.M. *et al.* 2004. Neurogenic inflammation in stress-induced termination of murine hair growth is promoted by nerve growth factor. *Am. J. Pathol.* **165:** 259–271.

32. Raychaudhuri, S.P., W.Y. Jiang & S.K. Raychaudhuri. 2008. Revisiting the Koebner phenomenon: role of NGF and its receptor system in the pathogenesis of psoriasis. *Am. J. Pathol.* **172:** 961–971.

33. Botchkarev, V.A. *et al.* 2006. Neurotrophins in skin biology and pathology. *J. Invest. Dermatol.* **126:** 1719–1727.

34. Santambrogio, L. *et al.* 1994. Nerve growth factor production by lymphocytes. *J. Immunol.* **153:** 4488–4495.

35. Raychaudhuri, S.K. & S.P. Raychaudhuri. 2009. NGF and its receptor system: a new dimension in the pathogenesis of psoriasis and psoriatic arthritis. *Ann. N.Y. Acad. Sci.* **1173:** 470–477.

36. Bracci-Laudiero, L. *et al.* 2005. Endogenous NGF regulates CGRP expression in human monocytes, and affects HLA-DR and CD86 expression and IL-10 production. *Blood* **106:** 3507–3514.

37. Nassenstein, C., O. Schulte-Herbruggen, H. Renz & A. Braun. 2006. Nerve growth factor: the central hub in the development of allergic asthma? *Eur. J. Pharmacol.* **533:** 195–206.

38. Peters, E.M. *et al.* 2011. Nerve growth factor partially recovers inflamed skin from stress-induced worsening in allergic inflammation. *J. Invest. Dermatol.* **131:** 735–743.

39. Groneberg, D.A. *et al.* 2007. Cell type-specific regulation of brain-derived neurotrophic factor in states of allergic inflammation. *Clin. Exp. Allergy* **37:** 1386–1391.

40. Truzzi, F. *et al.* 2011. p75 neurotrophin receptor mediates apoptosis in transit-amplifying cells and its overexpression restores cell death in psoriatic keratinocytes. *Cell Death Differ.* **18:** 948–958.

41. Namura, K. *et al.* 2007. Relationship of serum brain-derived neurotrophic factor level with other markers of disease severity in patients with atopic dermatitis. *Clin. Immunol.* **122:** 181–186.

42. Joachim, R.A. *et al.* 2008. Correlation between immune and neuronal parameters and stress perception in allergic asthmatics. *Clin. Exp. Allergy* **38:** 283–290.

43. Schulte-Herbruggen, O. *et al.* 2005. Tumor necrosis factor-alpha and interleukin-6 regulate secretion of brain-derived neurotrophic factor in human monocytes. *J. Neuroimmunol.* **160:** 204–209.

44. Besser, M. & R. Wank. 1999. Cutting edge: clonally restricted production of the neurotrophins brain-derived neurotrophic factor and neurotrophin-3 mRNA by human immune cells and Th1/Th2-polarized expression of their receptors. *J. Immunol.* **162:** 6303–6306.

45. Solomon, G.F. 1987. Psychoneuroimmunology: interactions between central nervous system and immune system. *J. Neurosci. Res.* **18:** 1–9.

46. Tausk, F., I. Elenkov & J. Moynihan. 2008. Psychoneuroimmunology. *Derm. Ther.* **21:** 22–31.

47. Seiffert, K. & R.D. Granstein. 2006. Neuroendocrine regulation of skin dendritic cells. *Ann. N.Y. Acad. Sci.* **1088:** 195–206.

48. Niizeki, H., I. Kurimoto & J.W. Streilein. 1999. A substance p agonist acts as an adjuvant to promote hapten-specific skin immunity. *J. Invest. Dermatol.* **112:** 437–442.

49. Nakano, Y. 2004. Stress-induced modulation of skin immune function: two types of antigen-presenting cells in the epidermis are differentially regulated by chronic stress. *Br. J. Dermatol.* **151:** 50–64.

50. Joachim, R.A. *et al.* 2008. Stress-induced neurogenic inflammation in murine skin skews dendritic cells towards maturation and migration: key role of intercellular adhesion molecule-1/leukocyte function-associated antigen interactions. *Am. J. Pathol.* **173:** 1379–1388.

51. Jessop, D.S., D. Renshaw, P.J. Larsen, *et al.* 2000. Substance P is involved in terminating the hypothalamo–pituitary–adrenal axis response to acute stress through centrally located neurokinin-1 receptors. *Stress* **3:** 209–220.

52. Renoldi, G. & R.W. Invernizzi. 2006. Blockade of tachykinin NK1 receptors attenuates stress-induced rise of extracellular noradrenaline and dopamine in the rat and gerbil medial prefrontal cortex. *J. Neurosci. Res.* **84:** 961–968.

53. Ebner, K. & N. Singewald. 2007. Stress-induced release of substance P in the locus coeruleus modulates cortical noradrenaline release. *Naunyn Schmiedebergs Arch. Pharmacol.* **376:** 73–82.

54. Li, Y. *et al.* 2009. Effects of unpredictable chronic stress on behavior and brain-derived neurotrophic factor expression in CA3 subfield and dentate gyrus of the hippocampus in different aged rats. *Chin. Med. J.* **122:** 1564–1569.

55. Cao, L. *et al.* 2010. Environmental and genetic activation of a brain-adipocyte BDNF/leptin axis causes cancer remission and inhibition. *Cell* **142:** 52–64.

56. Peters, E.M., P.C. Arck & R. Paus. 2006. Hair growth inhibition by psychoemotional stress: a mouse model for neural mechanisms in hair growth control. *Exp. Dermatol.* **15:** 1–13.

57. Siebenhaar, F. *et al.* 2007. Substance P as an immunomodulatory neuropeptide in a mouse model for autoimmune hair loss (Alopecia Areata). *J. Invest. Dermatol.* **127:** 1489–1497.

58. Ito, N., T. Ito & R. Paus. 2005. The human hair follicle has established a fully functional peripheral equivalent of the hypothalamic-pituitary-adrenal-axis (HPA). *Exp. Dermatol.* **14:** 158.

59. Peters, E.M.J. *et al.* 2007. Probing the effects of stress mediators on the human hair follicle: substance P holds central position. *Am. J. Pathol.* **71:** 1872–1886.

60. Novak, N. & T. Bieber. 2004. Pathophysiologie der atopischen Dermatitis. *Deutsches Ärzteblatt* **101:** 94–102.

61. Buske-Kirschbaum, A., M. Ebrecht & D.H. Hellhammer. 2010. Blunted HPA axis responsiveness to stress in atopic patients is associated with the acuity and severeness of allergic inflammation. *Brain Behav. Immun.* **24:** 1347–1353.

62. Buske-Kirschbaum, A. & D.H. Hellhammer. 2003. Endocrine and immune responses to stress in chronic inflammatory skin disorders. *Ann. N.Y. Acad. Sci.* **992:** 231–240.

63. Joachim, R.A. *et al.* 2003. Stress enhances airway reactivity and airway inflammation in an animal model of allergic bronchial asthma. *Psychosom. Med.* **65:** 811–815.

64. Yamaoka, J. & S. Kawana. 2007. Rapid changes in substance P signaling and neutral endopeptidase induced by skin-scratching stimulation in mice. *J. Dermatol. Sci.* **48:** 123–132.

65. van der Kleij, H.P. *et al.* 2003. Functional expression of neurokinin 1 receptors on mast cells induced by IL-4 and stem cell factor. *J. Immunol.* **171:** 2074–2079.

66. Siebenhaar, F. *et al.* 2008. Mast cell-driven skin inflammation is impaired in the absence of sensory nerves. *J. Allergy. Clin. Immunol.* **121:** 955–961.

67. Pavlovic, S. *et al.* 2008. Further exploring the brain-skin connection: stress worsens dermatitis via substance P-dependent neurogenic inflammation in mice. *J. Invest. Dermatol.* **128:** 434–446.

68. Raap, U. & A. Kapp. 2005. Neuroimmunological findings in allergic skin diseases. *Curr. Op. Allerg. Clin. Immunol.* **5:** 419–424.

69. Werfel, T. *et al.* 2009. Atopic dermatitis: S2 guidelines. *J. Dtsch. Dermatol. Ges.* **7**(Suppl 1): S1–S46.

70. Gaston, L., J.C. Crombez, M. Lassonde, *et al.* 1991. Psychological stress and psoriasis: experimental and prospective correlational studies. *Acta Derm. Venereol.* **156:** 37–43.

71. Naldi, L. *et al.* 2005. Cigarette smoking, body mass index, and stressful life events as risk factors for psoriasis: results from an Italian case-control study. *J. Invest. Dermatol.* **125:** 61–67.

72. Remrod, C., S. Lonne-Rahm & K. Nordlind. 2007. Study of substance P and its receptor neurokinin-1 in psoriasis and their relation to chronic stress and pruritus. *Arch. Dermatol. Res.* **299:** 85–91.

73. Schmid-Ott, G. *et al.* 1998. Stress-induced endocrine and immunological changes in psoriasis patients and healthy controls. A preliminary study. *Psychother. Psychosom.* **67:** 37–42.

74. Buske-Kirschbaum, A., M. Ebrecht, S. Kern & D.H. Hellhammer. 2006. Endocrine stress responses in TH1-mediated chronic inflammatory skin disease (psoriasis vulgaris)—do they parallel stress-induced endocrine changes in TH2-mediated inflammatory dermatoses (atopic dermatitis)? *Psychoneuroendocrinology* **31:** 439–446.

75. Sabat, R. *et al.* 2007. Immunopathogenesis of psoriasis. *Exp. Dermatol.* **16:** 779–798.

76. Bradley, J.R. 2008. TNF-mediated inflammatory disease. *J. Pathol.* **214:** 149–160.

77. Zachariae, R., C.O. Zachariae, U. Lei & A.F. Pedersen. 2008. Affective and sensory dimensions of pruritus severity:

associations with psychological symptoms and quality of life in psoriasis patients. *Acta Derm. Venereol.* **88:** 121–127.

78. Harvima, I.T. *et al.* 1993. Association of cutaneous mast cells and sensory nerves with psychic stress in psoriasis. *Psychother. Psychsom.* **60:** 168–176.

79. Jiang, W.Y., S.P. Raychaudhuri & E.M. Farber. 1998. Double-labeled immunofluorescence study of cutaneous nerves in psoriasis. *Int. J. Dermatol.* **37:** 572–574.

80. Ansel, J.C., J.R. Brown, D.G. Payan & M.A. Brown. 1993. Substance P selectively activates TNF-alpha gene expression in murine mast cells. *J. Immunol.* **150:** 4478–4485.

81. Okabe, T., M. Hide, O. Koro & S. Yamamoto. 2000. Substance P induces tumor necrosis factor-alpha release from human skin via mitogen-activated protein kinase. *Eur. J. Pharmacol.* **398:** 309–315.

82. Pincelli, C. 2000. Nerve growth factor and keratinocytes: a role in psoriasis. *Eur. J. Dermatol.* **10:** 85–90.

83. Al-Shakhli, H., D. Harcourt & J. Kenealy. 2006. Psychological distress surrounding diagnosis of malignant and non-malignant skin lesions at a pigmented lesion clinic. *J. Plast. Reconstr. Aesthet. Surg.* **59:** 479–486.

84. Fawzy, F.I., A.L. Canada & N.W. Fawzy. 2003. Malignant melanoma: effects of a brief, structured psychiatric intervention on survival and recurrence at 10-year follow-up. *Arch. Gen. Psych.* **60:** 100–103.

85. Boesen, E.H. *et al.* 2007. Survival after a psychoeducational intervention for patients with cutaneous malignant melanoma: a replication study. *J. Clin. Oncol.* **25:** 5698–5703.

86. Boesen, E., S. Boesen, S. Christensen & C. Johansen. 2007. Comparison of participants and non-participants in a randomized psychosocial intervention study among patients with malignant melanoma. *Psychosomatics* **48:** 510–516.

87. Ben-Eliyahu, S., G.G. Page & S.J. Schleifer. 2007. Stress, NK cells, and cancer: still a promissory note. *Brain Behav. Immun.* **21:** 881–887.

88. Goldfarb, Y. *et al.* 2011. Improving postoperative immune status and resistance to cancer metastasis: a combined perioperative approach of immunostimulation and prevention of excessive surgical stress responses. *Ann. Surg.* **253:** 798–810.

89. Glasner, A. *et al.* 2010. Improving survival rates in two models of spontaneous postoperative metastasis in mice by combined administration of a beta-adrenergic antagonist and a cyclooxygenase-2 inhibitor. *J. Immunol.* **184:** 2449–2457.

90. Steinkraus, V., M. Nose, H. Mensing & C. Korner. 1990. Radioligand binding characteristics of beta 2-adrenoceptors of cultured melanoma cells. *Br. J. Dermatol.* **123:** 163–170.

91. Reiche, E.M., S.O. Nunes & H.K. Morimoto. 2004. Stress, depression, the immune system, and cancer. *Lancet Oncol.* **5:** 617–625.

92. Saul, A.N. *et al.* 2005. Chronic stress and susceptibility to skin cancer. *J. Natl. Cancer Inst.* **97:** 1760–1767.

93. Dhabhar, F.S. 2008. Enhancing versus suppressive effects of stress on immune function: implications for immunoprotection versus immunopathology. *Allergy Asthma Clin. Immunol.* **4:** 2–11.

94. Dhabhar, F.S. *et al.* 2010. Short-term stress enhances cellular immunity and increases early resistance to squamous cell carcinoma. *Brain Behav. Immun.* **24:** 127–137.

95. Fischer, A. *et al.* 2005. Neuronal plasticity in persistent perennial allergic rhinitis. *J. Occup. Environ. Med.* **47:** 20–25.

Ann. N.Y. Acad. Sci. ISSN 0077-8923

17α-androstenediol-mediated oncophagy of tumor cells by different mechanisms is determined by the target tumor

Roger M. Loria[1,2] and Martin R. Graf[1]

[1]Department of Microbiology and Immunology, Virginia Commonwealth University Medical Center, Richmond, Virginia.
[2]Commonwealth University Reanimation, Engineering Science Center (VCURES), Richmond, Virginia

Address for correspondence: Roger M. Loria, Professor of Microbiology, Immunology, Pathology and Emergency Medicine, Virginia Commonwealth University Medical Center, 1101 E. Marshall Street, Richmond, VA 23298-0678. loria@vcu.edu

Δ5-androstene-3β,17α-diol (17α-AED) mediates oncophagy of human myeloid, glioma, and breast tumor cells by apoptotic- and autophagic-programmed cell death pathways, whereas the 17β-epimer does not. In hematologically derived myeloid tumor cells, 17α-AED induced apoptosis, as determined by TUNEL staining, caspase, PARP activation, and electron microscopy. In contrast, 17α-AED treatment of glioma cells of neuroectodermal lineaged induced autophagy, evident by the presence of acidic vesicular organelles, LC3 processing, and upregulation of beclin-1. Proliferation inhibition studies on primary and established glioma cells demonstrated that the IC-50 of the steroid is ~15 μM. In the case of breast cancer cells, the bioactivity of 17α–AED is independent of the expression of estrogen or androgen receptors. Collectively, oncophagy is induced by 17α-AED treatment in human tumor cells and proceeds by the induction of either autophagy or apoptosis. The neoplastic cell determines which oncophagic pathway is utilized.

Keywords: androstenediol; apoptosis; autophagy; glioblastoma; oncophagy

Introduction

Cellular death by a variety of processes, such as necrosis, apoptosis, and autophagy, can be considered a distinct phase of host-resistance processes in response to pathological conditions such as cancer, acute infections, and trauma.[1,2] We have reported that Δ5-androstene hormones—dehydroepiandrosterone, Δ5-androstene-3β,17β-diol (17β-AED), and Δ5-androstene-3β,7β,17β-triol (7β,17β-AET)—upregulate host resistance against lethal infections by bacteria and viruses, restore host myelopoiesis following γ- and α-whole body radiation exposure, and modulate shock response following severe hemorrhagic shock.[3–12] In contrast, when the hydroxyl group on C-17 is in the α-position (17α-AED, the epimer of 17β-AED), the steroid mediates host resistance by the induction of oncophagy of myeloid, breast, and glioma tumor cells by apoptotic or autophagic mechanisms.[13–15] It is possible that Δ5-androstenediol epimers may serve as an *in vivo* metabolic regulation point that

determines the direction of host resistance functions, as illustrated in Figure 1.

In our early investigations, we assessed antineoplastic activity of androstene steroids on human and murine HL-60, RAW264.7, and P388D1 myeloid tumor cells. In proliferation studies, adding increasing concentrations of 17α-AED (6.25–100 nM) to P388D1 for 48 h or HL-60 cells for 24 h significantly suppressed the incorporation of H^3 thymidine in a dose-dependent manner.[13] At 25 nM, 17α AED suppressed the proliferation of P388D1 or HL-60 cells by 53.89% and 83.15%, respectively ($P < 0.001$). HL-60 cells were more sensitive to the effects of 17α-AED.[13] It is relevant to emphasize that the epimer 17β AED had no inhibitory effect on proliferation or DNA synthesis, as measured by thymidine uptake of human promyelocytic leukemia cells or untreated controls cells. Similar results were obtained with other hematological tumor cell lines.

In tests of the effects of 17α-AED on the estrogen receptor–positive human breast cancer cell line

doi: 10.1111/j.1749-6632.2012.06602.x
Ann. N.Y. Acad. Sci. 1262 (2012) 127–133 © 2012 New York Academy of Sciences.

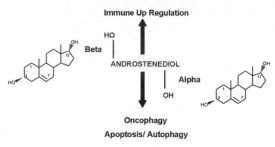

Figure 1. Immune regulation by β-androstenes and tumor cell death by α-androstenes. The figure illustrates the central role of androstenediol epimers in control of immune regulation and oncophagy.

ZR75-1 and the estrogen receptor–negative breast cancer cell line MBA-MB231, in both cell lines proliferation was markedly inhibited. This inhibition was shown to be dose dependent, with a minimal inhibitory dose of 12.5 nM in both cell lines.[13] These results showed that the inhibitory action of l7α-AED on human mammary carcinomas appears to be independent of either the α-estrogen or the androgen receptor. Since it could not be blocked by either 100 nM tamoxifen or 1.0 μM flutamine, inhibitors of the estrogen and androgen receptors, respectively.[14] As expected, the addition of 1.0 nM of 17β-estradiol increased the proliferation of ZR75-1 breast cancer cells by 22.6% but, unexpectedly, potentiated the inhibitory effects of 17α-AED, resulting in 78.1% inhibition, compared with the vehicle and estrogen only ($P < 0.001$).[14] At a dose of 12.5 nM, 17α-AED, in the presence of 17β-estradiol, resulted in an increased inhibition of 55.6% compared with 17α-AED alone. This antiproliferative effect of the combined steroid treatment was dose dependent and demonstrates that estrogen potentiates the inhibitory effect of 17α-AED.

Glioblastomas are grade IV astrocytic neoplasms of glial lineage derived from embryonic neuro-ectodermal tissue.[15] These primary central nervous system tumors are essentially incurable, with a median survival of approximately 18 months after primary diagnosis, regardless of treatment.[16–18] Based on our findings of the anticancer activity of 17α-AED on myeloid and breast tumor cells, we tested the effects of the steroid epimers on human TG98 glioma cell proliferation. 17α-AED inhibited more than 90% of proliferation within three days of treatment, while the β-epimer had a neglible inhibitory effect (Fig. 2). We expanded our studies to include several unrelated primary and established human

glioblastoma cell lines and evaluated the effects of 17α-AED on proliferation and viability. The average IC-50 for 17α-AED on these glioma cells was determined to be 14.28 μM ± 5.25. These findings established that 17α-AED can induce a significant level of cell death in a dose-dependent fashion in several different, unrelated malignant gliomas, whereas the β-epimer possessed only negligible antitumor activity.[19]

We next explored the structure–activity relation of several Δ5-androstene epimers, all with the C-3 hydroxyl group in the β-orientation, but with a different hydroxyl position at C-7 and/or C-17. The molecules studied were Δ5-androstene-3β,17α -diol (17α-AED); Δ5-androstene-3β,17β -diol (17β-AED); Δ5-androstene-3β,7α,17α -triol (7α,17α-AET); and Δ5-androstene-3β,7β,17α-triol (7β,17α AET). The results, shown in Figure 2, clearly indicate that the 17α-AED androstene derivative is the most potent inhibitor of glioma cell proliferation, whereas the β-epimer was relatively ineffective. Interestingly, the addition of a third hydroxyl group to C-7 in either the α- or β-orientation to 17α-AED decreased the antiproliferative activity of the steroid on tumor cells. This was unexpected because our previous findings on the immunoregulatory action of 17β-AED showed that the addition of the third hydroxyl group β-configuration to C-7 resulted in remarkable enhancement in host immune resistance to infections, compared with 17β-AED.[20] The ranking of the antitumor activities of these steroids and their IC50 on T98G glioma cells is 17α-AED (~10 μM) > 7α,17α-AET (~30 μM) ≫ 17β-AED (not achievable) ≈ 7β,17α-AET (not achievable). These findings show a clear structure–activity relation with the α-position of the androstene hydroxyl group at C-17. This position and configuration dictates the antitumor activity, while the hydroxyl group at the C-7 position, in either α- or β-position, is not critical for antitumor activity. Furthermore, androstene molecules with the hydroxyl groups in the beta configuration at C-3, C-7, and/or C-17 do not have significant antitumor activity. This specific antitumor activity of 17α-AED is evident in both glioma and leukemia tumor cells, while androstenes with hydroxyl on C-17 in the β-configuration do not have such activity.[20]

In order to gain insight into the mechanisms of the anticancer activity of 17α-AED on tumor

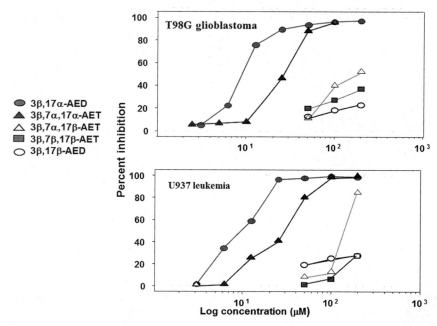

Figure 2. Inhibition of tumor cell proliferation by androstene steroids. (A) Human T98G glioblastoma or (B) U937 lymphoma cells were cultured with titrated doses of 3β,17α-AED (●); 3β,7α,17α-AET (▲); 3β,7α,17β-AET (△); 3β,7β,17β-AET (■); or 3β,17β-AED (o) in a three-day proliferation assay. Incorporation of 3H-TdR was used as a measure of cell proliferation, and the percentage in which proliferation was inhibited compared with sham-treated cultures is indicated. This is a representation of at least two to three independent experiments. 3β,17α-AED (●) is a more potent inhibitor of tumor cell proliferation than 3β,7α,17α-AET (▲) or the epimer 3β,7α,17β-AET (△). Compounds with a 17β hydroxyl configuration had significantly reduced antitumor proliferation activity. Adapted from Ref. 20.

cells, we investigated the activation of programmed cell death pathways in treated leukemic and glioma cells. In the first series of studies, we analyzed 17α-AED–treated myeloid tumor and glioma cells for the induction of apoptotic cell death. The number of tumor cells exposed to 17α-AED or vehicle undergoing apoptosis was identified by the presence of double-stranded DNA fragmentation and quantified after fluorescein isothiocyanate-labeled dUTP staining (TUNEL reaction, Boehringer Mannheim) and flow cytometry. A minimum of 10,000 events were collected for each sample. For negative and positive control of apoptosis, cells were incubated in the absence of terminal transferase or in the presence of 2.5 μg of camptothecin, an inhibitor of the DNA enzyme topoisomerase I (positive control).[21] Necrotic and late-stage apoptotic cells were determined by the uptake of propidium iodide in parallel treated cultures. The results shown in Figure 3A indicate that 54.1% of the U937 lymphoma cultures treated with 17α-AED cells were TUNEL positive, demonstrating that these cells were

apoptotic. Furthermore, the induction of apoptosis by 17α-AED treatment of U937 lymphoma cells was dose dependent, as revealed by annexing V staining, with an LD50 between 6.25 μM and 12.5 μM. No appreciable evidence of apoptosis was detected by TUNEL or annexin V staining in U937 lymphoma cells treated with 17β-AED or vehicle. In contrast to our findings with leukemic cells, no significant degree of DNA fragmentation could be detected in similar TUNEL/flow cytometric experiments in human U251MG and T98G glioblastoma cells treated with 17α-AED (Fig. 3A).[19]

To complement the TUNEL assays, we performed electron microscopy. RAW 264.7 leukemic cells and T98G glioma cells were treated with 17α-AED or vehicle. Mouse leukemic monocytes cells were exposed to 50 nM of the steroid for 48 h and then processed for electron microscopy. In Figure 3A, clear cell membrane blebbing, cell shrinkage, margination of chromatin, and formation of apoptotic bodies, all classical morphological characteristics of apoptosis, are readily detectable in the treated RAW

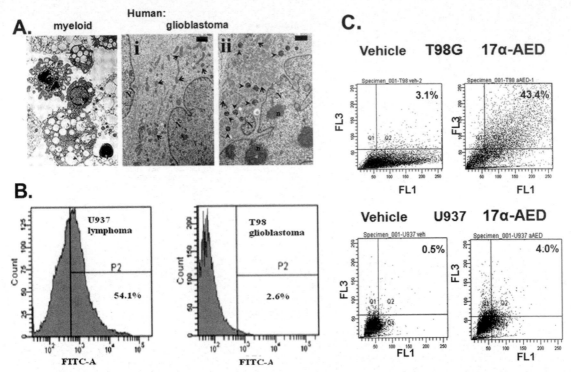

Figure 3. Analysis of the ultrastructure of human myeloid and glioblastoma cells treated with 17α-AED. (**A**) Ultrastructure analysis of mouse myeloid and human glioblastoma cells treated with 17α-AED. Mouse leukemic monocyte macrophage cell line. RAW 264.7 cells were exposed to 17α-AED (50 nM) for 48 hours. Degradation of the nuclearmembrane, blebbing of the cellular membrane and DNA condensation/fragmentation that are morphological features of apoptosis are observable (4600X). Untreated normal control not shown. T98G malignant glioma cells treated for three days with (i) vehicle or (ii) 17α-AED (10 μM). 17α-AED–treated T98G glioma cells with multiple autolysosomelike vesicles in the cytoplasm (arrows). Note that the nucleus (N) in control and treated T98G cells was intact with no evidence of chromatin condensation or fragmentation. (iii) Boxed region of 17α-AED-treated T98G cells shown in (ii) at a highermagnification. Bar = 1 μm. Adapted from Refs. 13 and 22. (**B**) α-AED treatment induces apoptotic DNA fragmentation of U937 lymphoma cells but not in T98G glioma cells. U937 lymphoma cells and T98G glioma cells. After four days, the degree of DNA cleavage was assessed by TUNEL staining and used as a measure of apoptosis. Annotated numbers reflect the percentage of cells staining positive for apoptosis (x-axis, FITC-A). This experiment was performed twice and yielded similar results. Adapted from Ref. 19. (**C**) Acridine orange staining of 17α-AED–treated tumor cells indicates a high level of acidic vacuoles in T98G glioma cells, but not in U937 lymphoma cells. (i) Treatment with 10 μM of 17α-AED induces a high level of AVO formation in T98G cells. A low level of acidic vacuoles can be detected in U937 cells treated with 10 μM 17α-AED (right) compared to vehicle treated cells (left). Annotated numbers indicate the percentage of cells positive for acidic vacuoles, quadrant Q2. Representative results of at least three independent experiments are shown. Adapted from Ref. 19.

264.7 cells.[13] Such morphological changes were not evident in RAW 264.7 cells following treatment either with the epimer 17β-AED or by the drug vehicle. In contrast, no ultrastructural features consistent with apoptotic cell death could be detected in T98G glioma cells treated with 17α-AED (10 μM for three days) upon electron microscopic analysis (Fig. 3A).[22] This lack of morphological evidence for apoptosis in the treated glioma cells is consistent with the TUNEL results. Upon closer examination of the ultrastructural features of 17α-AED–treated glioma cells, we detected the presence of multiple,

autolysosomal-like bodies in the cytoplasm of the T98G cells that were rarely present in the control-treated T98G cells (Fig. 3B). The vesicles tended to have the resemblance of late-stage autophagosomes or autolysosomes, were typically located adjacent to the nucleus, and appeared as dark vesicular bodies.[23,24]

During the late stages of autophagy, autophagosomes fuse with lysosomes, and the protein contents are degraded in the acidic environment of the organelle. The presence of these acidic vesicular organelles (AVOs) is a characteristic of

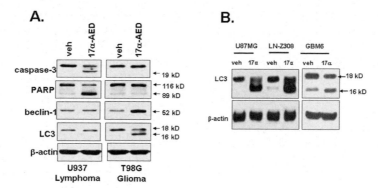

Figure 4. Treatment with 17α-AED induces apoptosis in U937 lymphoma cells and autophagy in T98G glioblastoma cells. (A) Whole-cell lysates were obtained from U937 and T98G tumor cells and treated for 48 h with an equivalent amount of vehicle (veh) or with 17α-AED (10 μM), inducing apoptosis in U937 lymphoma cells and autophagy in T98G glioblastoma cells. Adapted from Ref. 2. (B) A total of 20 μg of total protein was subjected to Western blot analysis and probed with antibodies specific for caspase-3, PARP, beclin-1, or LC-3. Equal protein loading is shown by immunoblotting for β-actin. This experiment was performed twice and yielded similar results. (B) 17α-AED treatment induces LC3 cleavage in unrelated human malignant glioma cell lines and primary human glioblastoma cells. Cell lysates were obtained from human glioblastoma U87MG, LN-Z308, and primary GBM6 glioma cells treated with 17α-AED (10 μM) or vehicle for 48 hours. A total of 25 μg of total protein was used for Western blot analysis using anti-LC3 antibody. LC3-II protein is readily detected in all of the steroid treated glioma cells and is absent or barely detectable in cells treated with vehicle. An anti-β–actin mAb was used to show equal protein loading.

autophagy and can be detected by acridine orange staining and quantified by flow cytometry.[25] We treated T98G glioma cells and U937 lymphoma cells with 17α-AED (10 μM, 3 days) and processed the cells for AVO detection. More than 35% of T98G glioma cells were positive for AVOs; in contrast, only 4.0% of U937 lymphoma cells treated with 17α-AED–contained AVOs (Fig. 3C). A high percentage of AVOs was also detected in U251MG malignant glioma cells treated with 17α-AED as compared with sham-treated cells (data not shown). Bafilomycin A1 specifically inhibits AVO formation by blocking the fusion of autophagosomes and lysosomes.[26] 3-Methyladenine is an inhibitor of class III PI3K and blocks early-stage, autophagic sequestration.[27] The addition of bafilomycin A1 or 3-methyladenine to T98G cultures treated with 17α-AED, either completely or significantly, blocked AVO formation.[22] These results suggest that 17α-AED induces the formation of AVOs, which is a feature of autophagy, in human glioma cells but not in lymphoma cells.

Biochemical analysis was performed on U937 lymphoma and T98G glioma cells exposed to 17α-AED to more definitively resolve the process of oncophagy in these tumor cells of unrelated embryonic lineages and tissue origins. Tumor cells were treated with 10 μM 17α-AED, a dose corresponding to their

~LD50.[22] After 48 h, tumor cells were harvested and analyzed for biochemical markers specific for the induction of apoptosis or autophagy. Proteolytic processing of caspase-3 and poly ADP (adenosine diphosphate)-ribose polymerase (PARP) cleavage are well-characterized events indicative of the activation of the intrinsic and/or extrinsic apoptotic pathway.[28,29] Increased levels of beclin-1, a protein required for the initiation of the autophagic process, and the cleavage of microtubule-associated protein-light chain 3 (LC3)-I to LC3-II, an essential step in autophagosome membrane formation and extension, are biochemical markers for the induction of autophagy.[30] Elevated levels of caspase-3 cleavage and PARP processing were detected in U937 lymphoma cells treated with 17α-AED in the absence of the conversion of LC3-I (18 kD) to LC3-II (16 kD) or beclin-1 upregulation (Fig. 4A). In contrast, increased protein levels of beclin-1 and the conversion of LC3-I to LC3-II were readily apparent in T98G glioma cells treated with 17α-AED, without increased activation of caspase-3 and PARP, compared with sham-treated T98G cells (Fig. 4A). We extended our investigations to determine whether 17α-AED would induce autophagy in other glioma cell lines and primary human glioblastoma cells. In these studies, U87MG, LN-Z308, and primary GBM6 glioma cells were exposed to 17α-AED or vehicle and LC3 conversion was assessed. The results in Figure 4B show that there is robust LC3 cleavage in all of the glioma cells treated with 17α-AED compared with controls. Additionally, we have reported that disruption of the autophagic pathway by *ATG5*- or *beclin-1*–targeted siRNA blocks the antitumor activity of 17α-AED on human glioma cells.[22,31] These results indicate that treatment with 17α-AED induces oncophagy in hematological cancer cells by the activation of the apoptotic pathway and the autophagic pathway in human glial tumor cells.

Discussion

These results demonstrate that 17α-AED treatment induces human hematological cancer cells, such as HL-60 and U937 myelocytic cells, to undergo apoptosis, which is in accordance with our previous reports.[13] In direct contrast, the cytotoxic activity of 17α-AED on human malignant glioma cells is mediated by autophagy, type 2-programmed cell death, as opposed to the classical apoptotic pathway. These findings strongly suggest that 17α-AED induces oncophagy in tumor cells by different programmed cell death pathways that may be determined by the germ line origin of the target tumor tissue. Hematological cancers, whose tissue is derived from the mesoderm layer, undergo apoptosis when treated with 17α-AED. On the other hand, tumors of glial lineage, which arise from the ectodermal layer during histogenesis, undergo autophagy when exposed to the same steroid agent.[19,20,31] We have recently reported that 17α-AED–induced autophagic cell death in human glioma cells is dependent on the activation of the endoplasmic reticulum stress pathway and early protein kinase-like endoplasmic reticulum kinase/eukaryotic translational initiation factor 2α signaling.[31] Human MDA-MB231 breast adenocarcinoma and ZR75-1 mammary ductal carcinoma cells, and other unrelated breast cancer cells, undergo cell death when exposed to 17α-AED. Unlike many glands, the mammary gland is derived from the ectodermal layer and is supported by connective tissue from the mesenchyme.[32] Preliminary studies from our laboratory indicate that 17α-AED induces autophagic cell death in human breast cancer cells (manuscript in preparation).

Structure–activity investigations revealed that the α-position of the hydroxyl group on C-17 of Δ5-androstene epimers is the major functional determinant for antitumor activity of the androstenes we analyzed. The addition of a third hydroxyl in the α-position on C-7 to form 7α,17α-AET reduced the potency of 17α-AED, although both androstenes induce autophagic cell death in glial tumors and apoptosis in leukemic cells.

Our studies on the tumor effects of androstene derivatives demonstrate for the first time that the same agonist, in this case 17α-AED, mediates the induction of two distinctly different program cell death pathways in tumor target cells originating from different histological origins. Consequently, it is apparent that the tumor tissue of origin, and possibly the germ layer, of the cancer treated with 17α-AED may determine the oncophagic pathway utilized.

Conflicts of interest

The authors declare no conflicts of interest.

References

1. Deretic, V. & B. Levine. 2009. Autophagy, immunity, and microbial adaptations. *Cell Host Microbe* **5:** 527–549.
2. Levine, B. & V. Deretic. 2007. Unveiling the roles of autophagy in innate and adaptive immunity. *Nat. Rev. Immunol.* **7:** 767–777.
3. Loria, R.M., T.H. Inge, S. Cook, *et al.* 1988. Protection against acute lethal viral infections with the native steroid dehydroepiandrosterone (DHEA). *J. Med. Virol.* **26:** 301–314.
4. Loria, R.M. & D.A. Padgett. 1991. Androstenediol regulates systemic resistance against lethal infections in mice. *Arch. Virol.* **127:** 103–115.
5. Padgett, D.A., R.M. Loria & J.F. Sheridan. 1997. Endocrine regulation of the immune response to influenza virus infection with a metabolite of DHEA—androstenediol. *J. Neuroimmunol.* **78:** 203–211.
6. Padgett, D.A. & R.M. Loria. 1998. Endocrine regulation of murine macrophage function: effects of dehydroepiandrosterone, androstenediol, and androstenetriol. *J. Neuroimmunol.* **84:** 61–68.
7. Ben-Nathan, D., D.A. Padgett & R.M. Loria. 1999. Androstenediol and dehydroepiandrosterone protect mice against lethal bacterial infections and LPS toxicity. *J. Med. Microbiol.* **48:** 425–431.
8. Padgett, D.A. & R.M. Loria. 1994. In vitro potentiation of lymphocyte activation by dehydroepiandrosterone, androstenediol, and androstenetriol. *J. Immunol.* **153:** 1544–1552.
9. Loria, R.M. & D.A. Padgett. 1993. Androstenediol regulate systemic resistance against lethal infections in mice. *N.Y. Acad. Sci.* **685:** 293–296.
10. Loria, R.M., D.H. Conrad, H. Huff & D. Ben-Nathan. 2000. Androstenetriol and androstenediol protect against lethal

radiation and restore radiation mediated immune injury. *N.Y. Acad. Sci.* **917:** 860–868.

11. Marcu, A.C., A.D. Kielar, K.F. Paccionea, *et al.* 2006. Androstenetriol improves survival in a rodent model of traumatic shock. *Resuscitation* **71:** 379–386.

12. Marcu, A.C., K.F. Paccionea, W.R Barbee, *et al.* 2007. Androstenetriol immunomodulation improves survival severe trauma-hemorrhage shock model. *J. Trauma Injury Infect. Crit. Care* **63:** 662–699.

13. Huynh, P.N. & R.M. Loria. 1997. Contrasting effects of alpha and beta androstenediol on oncogenic myeloid cell lines in vitro. *J. Leukocyte Biol.* **62:** 258–267.

14. Huynh, P.N., W.H. Carter & R.M. Loria. 2000. 17 Androstenediol inhibitory function is independent of estrogen receptor. *Cancer Detect. Prevent.* **24:** 435–445.

15. Field, M., A. Alvarez, S. Bushnev & K. Sugaya. 2010. Embryonic stem cell markers distinguishing cancer stem cells from normal human neuronal stem cell populations in malignant glioma patients. *Clin. Neurosurg.* **57:** 151–159.

16. Surawicz, T.S., F. Davis, S. Freels, *et al.* 1998. Brain tumor survival: results from the National Cancer Data Base. *J. Neurooncol.* **40:** 151–160.

17. Ohgaki, H. & P. Kleihues. 2007. Genetic pathways to primary and secondary glioblastoma. *Am. J. Pathol.* **170:** 1445–1453.

18. Mischel, P.S. & T.F. Cloughesy. 2003. Targeted molecular therapy of gbm. *Brain Pathol.* **13:** 52–61.

19. Graf, M.R., W. Jia & R.M. Lori. 2007. The neuro-steroid, 3 androstene 17 diol exhibits potent cytotoxic effects on human malignant glioma and lymphoma cells through different programmed cell death pathways. *Br. J. Cancer.* **97:** 619–627.

20. Graf, M.R., W. Jia, M.L. Lewbart & R.M. Loria. 2009. The anti-tumor effects of androstene steroids exhibit a strict structure-activity relationship dependent upon the orientation of the hydroxyl group on carbon-17. *Chem. Biol. Drug Design* **74:** 625–629.

21. Wall, M.E., M.C. Wani, C.E. Cook, *et al.* 1966. Plant antitumor agents: I. The isolation and structure of camptothecin, a novel alkaloidal leukemia and tumor inhibitor from camptotheca acuminate. *J. Am. Chem. Soc.* **88:** 3888–3890.

22. Graf, M.R., W. Jia, R.S. Johnson, *et al.* 2009. Autophagy and the functional roles of Atg5 and beclin-1 in the anti-tumor effects of 3 alpha androstene 17alpha diol neuro-steroid on malignant glioma cells. *J. Steroid Biochem. Mol. Biol.* **115:** 137–143.

23. Kroemer, G. & B. Levine. 2008. Autophagic cell death: the story of a misnomer. *Nat. Rev. Mol. Cell Biol.* **12:** 1004–1010.

24. Yi, J. & X.M. Tang. 1999. The convergent point of the endocytic and autophagic pathways in leydig cells. *Cell Res.* **4:** 243–253.

25. Paglin, S., T. Hollister, T. Delohery, *et al.* 2001. A novel response of cancer cells to radiation involves autophagy and formation of acidic vesicles. *Cancer Res.* **15:** 439–444.

26. Yamamoto, A., Y. Tagawa, T. Yoshimori, *et al.* 1998. Bafilomycin A1 prevents maturation of autophagic vacuoles by inhibiting fusion between autophagosomes and lysosomes in rat hepatoma cell line, H-4-II-E cells. *Cell Struct. Funct.* **23:** 33–42.

27. Gagliardi, S., M. Rees & C. Farina. 1999. Chemistry and structure activity relationships of bafilomycin A1, a potent and selective inhibitor of the vacuolar H+-ATPase. *Curr. Med. Chem.* **6:** 1197–212.

28. Takeuchi, H., Y. Kondo, K. Fujiwara, *et al.* 2005. Synergistic augmentation of rapamycin-induced autophagy in malignant glioma cells by phosphatidylinositol 3-kinase/protein kinase B inhibitors. *Cancer Res.* **15:** 3336–3346.

29. Boya, P., R.A. Gonzalez-Polo and N. Casares. 2005. Inhibition of macroautophagy triggers apoptosis. *Mol. Cell Biol.* **25:** 1025–1040.

30. Yousefi, S., R. Perozzo, I. Schmid *et al.* 2006. Calpain-mediated cleavage of Atg5 switches autophagy to apoptosis. *Nat. Cell Biol.* **8:** 1124–1132.

31. Jia, W., R.M. Loria, M.A. Park *et al.* 2010. The neuro-steroid, 5-androstene 3β,17α diol; induces endoplasmic reticulum stress and autophagy through PERK/eIF2α signaling in malignant glioma cells and transformed fibroblasts. *Int. J. Biochem. Cell Biol.* **42:** 2019–2029.

32. Cunha, G.R. 1994. Role of mesenchymal-epithelial interactions in normal and abnormal development of the mammary gland and prostate. *Cancer* **74**(Suppl): 1030–1044.

Ann. N.Y. Acad. Sci. ISSN 0077-8923

ANNALS OF THE NEW YORK ACADEMY OF SCIENCES

Issue: *Neuroimmunomodulation in Health and Disease*

Cellular and molecular players in the atherosclerotic plaque progression

Rita Businaro,[1] Angela Tagliani,[1] Brigitta Buttari,[2] Elisabetta Profumo,[2] Flora Ippoliti,[3] Claudio Di Cristofano,[1] Raffaele Capoano,[4] Bruno Salvati,[4] and Rachele Riganò[2]

[1]Department of Medico-Surgical Sciences and Biotechnologies, Sapienza University of Rome, Latina, Italy. [2]Department of Infectious, Parasitic, and Immune-Mediated Diseases, Istituto Superiore di Sanità, Rome, Italy. [3]Department of Experimental Medicine, Sapienza University of Rome, Rome, Italy. [4]Department of Surgical Sciences, Sapienza University of Rome, Rome, Italy

Address for correspondence: Rita Businaro, M.D., Ph.D., Department of Medico-Surgical Sciences and Biotechnologies, Sapienza University of Rome, Corso della Repubblica 79, 04100 Latina, Italy. rita.businaro@uniroma1.it

Atherosclerosis initiation and progression is controlled by inflammatory molecular and cellular mediators. Cells of innate immunity, stimulated by various endogenous molecules that have undergone a transformation following an oxidative stress or nonenzymatic glycation processes, activate cells of the adaptive immunity, found at the borders of atheromas. In this way, an immune response against endogenous modified antigens takes place and gives rise to chronic low-level inflammation leading to the slow development of complex atherosclerotic plaques. These lesions will occasionally ulcerate, thus ending with fatal clinical events. Plaque macrophages represent the majority of leukocytes in the atherosclerotic lesions, and their secretory activity, including proinflammatory cytokines and matrix-degrading proteases, may be related to the fragilization of the fibrous cap and then to the rupture of the plaque. A considerable amount of work is currently focused on the identification of locally released proinflammatory factors that influence the evolution of the plaque to an unstable phenotype. A better understanding of these molecular processes may contribute to new treatment strategies. Mediators released by the immune system and associated with the development of carotid atherosclerosis are discussed.

Keywords: atherosclerosis; inflammation; cytokines; macrophages

Morphogenesis of the atherosclerotic plaques

Atherosclerosis is a chronic progressive disease controlled at every stage, including initiation, progression, and thrombotic complications, by inflammatory cellular and molecular mediators.[1–3] When the endothelium is activated it expresses chemokines, including monocyte chemotactic protein 1 (MCP-1) and interleukin (IL)-8, and adhesion molecules, such as intercellular adhesion molecule 1 (ICAM-1), vascular adhesion molecule 1 (VCAM-1), and E- and P-selectin, which leads to monocyte/lymphocyte and polymorphonuclear granulocyte recruitment and infiltration into the subendothelium. Many risk factors traditionally associated with the onset of atherosclerosis, such as hyper-

tension and hypercholesterolemia, lead to an increased expression of adhesion molecules on endothelial cells, which promotes rolling and adhesion of mononuclear blood cells to vascular walls and their subsequent migration to subendothelial spaces by diapedesis. Monocytes differentiate into macrophages within tunica intima and expose on their surface scavenger receptors, thus binding and phagocytizing modified lipoproteins accumulated in the intima space. Lipoproteins infiltrate the artery wall where they undergo enzymatic modifications or nonenzymatic oxidation, thus triggering further monocyte and lymphocyte recruitment. Initially, lipid droplets accumulate in fatty streaks, trapped by the hyperelongated glycosaminoglycan chains on proteoglycans. As the amount of the lipids increases in fatty streaks,

doi: 10.1111/j.1749-6632.2012.06600.x

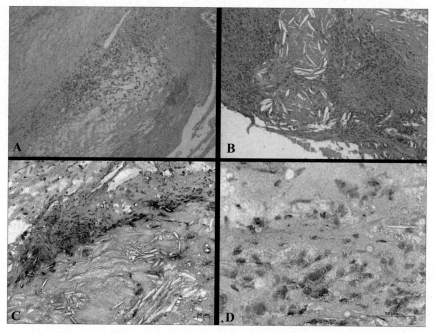

Figure 1. Human carotid plaque. (A) Hematoxylin-eosin staining: an inflammatory infiltrate at the border of the plaque is depicted (100×); (B) abundant inflammatory cells accumulate close to cholesterol crystals (200×); (C) CD68+ cells represent the main cell subtypes (200×); (D) a higher magnification (400×) allows the detection of a CD68 granular reaction in scattered cells.

macrophages infiltrate toward the deposited lipid to form pathological intima thickening with foam cells.[4,5] Oxidized low-density lipoproteins (oxLDL) also behave as a potent inflammatory agent: they stimulate the expression of adhesion molecules on endothelial cells, inducing endothelial dysfunction, have chemoattractant activity on monocytes, stimulate their maturation to macrophages, and inhibit their mobility. In addition, oxLDL binding to CD36 triggers the release of proinflammatory cytokines by macrophages. These cytokines, in turn, activate endothelial cells, smooth muscle cells, macrophages, and lymphocytes to produce proinflammatory cytokines that leads to a self-perpetuating inflammatory process that becomes less dependent on the presence of oxLDL. A number of factors found in the atherosclerotic plaque can participate in maintaining and amplifying cytokine production, giving rise to a cytokine network that may serve as a final common proinflammatory pathway, regardless of the initiating event, and may also provide a supplemental therapeutic target, especially in the late stages of the disease. Low-grade chronic inflammation is characterized by increased systemic levels of some cytokines and C-reactive protein, and a number of

studies have confirmed an association between low-grade systemic inflammation on the one hand and atherosclerosis and type 2 diabetes on the other.[5,6]

Inflammatory infiltrate subpopulations

Several studies were undertaken by our group to identify cell subpopulations and molecular mediators playing a central role in atherosclerotic plaque progression toward a stable or unstable phenotype.[7] We analyzed both human carotid plaques obtained by endarterectomy, and sera samples derived from the same patients. Our results showed that inflammatory infiltrates were particularly abundant in complicated plaques (Fig. 1A and B), whereas stable plaques exhibited a thick fibrous cap, lined with scattered inflammatory cells. The immunohistochemical characterization of cell subpopulations of the inflammatory infiltrate is depicted in Figures 1 and 2. The abundant presence of macrophages (CD68+ cells: Fig. 1C and D; CD14+ cells: Fig. 2B) both at the border of atheroma, in the close proximity of cholesterol crystals, and in the subendothelial space at the level of the fibrous cap, accounts for their prominent role in atherogenesis and plaque development.[8] Plaque macrophages represent the

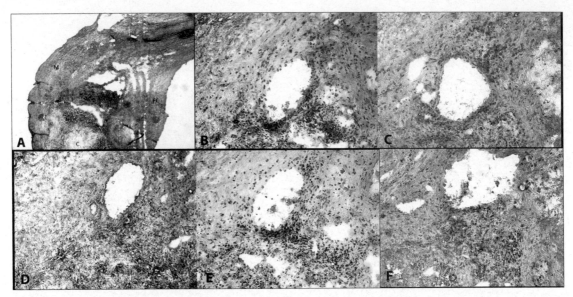

Figure 2. Human carotid plaque. (A) Hematoxylin-eosin staining (40×); (B) CD14[+] cells; (C) CD20[+] cells; (D) HSP90[+] cells; (E) CD8[+] cells; (F) CD3[+] cells (200×).

majority of leukocytes in the atherosclerotic lesions, and their secretory activity, including proinflammatory cytokines and matrix-degrading proteases, may be related to the fragilization and then rupture of the plaque, giving rise to atherothrombosis and thromboembolisms underlying the leading causes of death in western countries, such as myocardial infarction and stroke.[9] Chemokines are the most important chemotactic factors for recruiting inflammatory cells to atherosclerotic lesions, as demonstrated by the inhibition of CCR2, CX3CR1, and CCR5, which abrogates the development of atheroma in hypercholesterolemic mice.[10] Following chemokine expression, monocytes become tethered and roll on endothelial cells, firmly adhering through the interaction of monocyte integrins with endothelial cell ligands.[11] Several lines of evidence have recently led to a reconsideration of the role of polymorphonuclear cells in atherogenesis, initially underestimated for the low frequency of neutrophil detection in human and murine atherosclerotic lesions.[12] This low frequency is probably due to the short life span of these cells in the tissue, whereas peripheral neutrophil counts have been proven to closely correlate with atherosclerotic lesion size in mice, as well as with incidence of myocardial infarction in humans.[13] A recent systematic review showed that neutrophils were independent predictors of cardiovascular outcomes when analyzed concomitantly with other markers of inflammation.[14] In addition, statin treatment was shown to be consistently associated with the reversion of the dysregulated polymorphonuclear (PMN) function occurring during the preclinical development of atherosclerosis,[15] underscoring the potential importance of modulating neutrophilic inflammation as part of a novel strategy for preventing and treating atherosclerosis.[16]

Macrophage polarization

The inflammatory milieu characterizing the wall of medium and large arteries affected by atherosclerosis polarizes monocytes migrated from circulation toward an inflammatory phenotype that will then give rise to inflammatory macrophages and dendritic cells.[17,18] Indeed, mononuclear phagocytes assume different morphological and functional phenotypes in response to the tissue microenvironment,[19] and several studies suggest that different stages in the progression of atherosclerosis are associated with the presence of distinct macrophage subtypes: the classically activated (M1) and alternatively activated (M2) populations.[20] When a circulating monocyte leaves the bloodstream, following binding to the upregulated adhesion molecules expressed on dysfunctional endothelium, it penetrates into the subendothelial space, it acquires pro- or anti-inflammatory function and morphology, depending on the signals from the local microenvironment. In this sense, the cell becomes polarized toward a specific activation state, of which the phenotypes M1 and M2 represent two well-known types. T helper 1 (Th1) cytokines, including IFN-γ, TNF-α, and GM-CSF, are responsible for the polarization of macrophages toward a proinflammatory M1 phenotype; on the other hand, Th2 cytokines, including IL-4, IL-5, and IL-13, induce the M2 phenotype, which, generally, prevents tissue damage and increases phagocytosis of apoptotic cells and cellular debris (efferocytosis). Human M1 macrophages are characterized by an IL-12^{high}, IL-23^{high}, IL-10^{low} phenotype and are involved in the production of proinflammatory cytokines, such as IL-1β, IL-6, and TNF-α, as well as in the release of reactive oxygen species and nitrogen intermediates. M2 macrophages have an IL-12^{low}, IL-23^{low}, IL-10^{high} phenotype and express high levels of scavenger receptors. Some studies, taking advantage of the apoE-knockout (KO) mouse model, established that atherosclerotic early lesions are enriched with M2 macrophages, whereas more advanced lesions show an abundance of M1 cells. From these data, one may conclude that the evolution and complication of atherosclerosis is affected by the balance between M1 and M2 macrophage subpopulations.[21] Moreover, inflammatory cytokines stimulate the phagocytic functions of the proinflammatory M1 macrophages and their antigen-presenting capacity.

T cell subpopulations

Dendritic cells are immune cells that internalize, process, and present antigens, leading to activation or suppression of T cells.[11] In line with this, many experimental studies have demonstrated the presence of activated T cells in human atherosclerotic plaques, showing that adaptive immunity is deeply involved in the development of atherosclerosis.[22] Initial studies have focused on the role of Th1 and Th2 responses in atherosclerosis, and more recently, evidence has been published supporting a protective role of regulatory T cells (T_{reg} cells) in this disease. A third member of the T helper set, IL-17–producing Th17 cells had been described as a distinct lineage that plays an important role in autoimmune diseases.[23] Accordingly accelerated atherosclerosis has been reported in patients with various autoimmune diseases, suggesting an involvement of autoimmune mechanisms in atherogenesis.[24] Modified self-antigens have been detected within atherosclerotic lesions as well as humoral and cellular immune responses directed against self-antigens in atherosclerosis patients. Our studies, directed to identify the main cellular and molecular players responsible for the progression of atherosclerosis, have focused on the identification of putative autoantigens involved in atherosclerosis development.

Autoantigens and plaque development

Heat shock proteins

Endothelial cells subjected to various stress conditions, including oxidative stress, chemicals, viruses, and ischemia-reperfusion injury, express increased amounts of heat shock proteins (HSPs)—some of the most successfully conserved proteins throughout evolution. The increased expression of HSPs is observed not only within the cytoplasmic compartment, but also extends to the cell membrane, and leads to the release of HSPs at the level of intercellular space, where they may trigger an autoimmune response. In fact, intracellular HSPs have cytoprotective functions, whereas extracellular-located or membrane-bound HSPs mediate immunological functions.[25]

Our previous results showed that HSP90 is overexpressed in human carotid atherosclerotic plaque (Fig. 2D) and is present in a soluble form within the sera of patients. The sera also contained anti-HSP90 antibodies; and a cellular immune response against HSP90 was elicited following HSP90 stimulation of polyclonal population of T cells isolated from carotid plaques and from patients' peripheral blood mononuclear cells (PBMC).[26,27] Our observations suggest that HSP90, produced in large quantities during stress, may be neoexpressed on cellular membranes and released in the vascular flow, where it may interact with professional antigen-presenting cells through distinct surface receptors. Another member of the HSP90 family, gp96, binds to Toll-like receptor (TLR) 2/TLR4, signaling receptors capable of inducing cytokine release and amplifying an inflammatory response. Binding to cells may elicit the release of proinflammatory cytokines and the activation of an immune response against the protein itself, thus stimulating cellular and humoral immune responses.

Beta-2-glycoprotein 1

Another plasma protein that seems to deserve great attention because of its involvement in immune system activation during atherosclerosis progression is beta-2-glycoprotein 1, a plasma protein that functions as a phospholipid cofactor and a natural anticoagulant. Its presence was detected within atherosclerotic plaques, and we demonstrated that its oxidized as well as its glycated form is able to activate immature monocyte-derived dendritic cells. In addition, these dendritic cells, following stimulation, produce cytokine-targeting Th1 cells, starting in this way a Th1-type response, amplifying inflammatory processes underlying endothelial dysfunction.[28,29]

Hemoglobin

Human oxidized hemoglobin (Hb) was shown to be an antigenic target of cell-mediated immune reactions in carotid atherosclerosis.[30] Intraplaque hemorrhage is an important process in the evolution of plaques from a stable phenotype of high-risk unstable lesions. The source of red blood cells within the plaques has been identified in the leaky immature microvessels that are present around and within the plaque and frequently break, resulting in the deposition of erythrocytes and the release of large amounts of free Hb. Free Hb is subsequently cleared by the macrophage Hb scavenger receptor CD163,[31] whose expression is increased in acute coronary syndromes.[32] Preliminary findings showed that human Hb acts as a self-antigen in patients with carotid atherosclerosis.[30,33]

Moreover, we recently demonstrated that Hb, by interacting with CD163 on monocytes and immature dendritic cells, might induce cell recruitment and activation within the vascular wall, thus contributing to the complex cross-talk of chemotactic signals that mediate atherosclerotic lesion instability. The high frequency of dendritic cells observed within complicated plaques, characterized by intraplaque hemorrhage (Fig. 3), sustains the chemotactic role of hemoglobin released from red blood cells.[34]

The conclusion to be drawn from these data is that cells of innate immunity, stimulated by various endogenous molecules that have undergone a transformation following an oxidative stress or nonenzymatic glycation processes, activate cells of the adaptive immune system found at the bor-

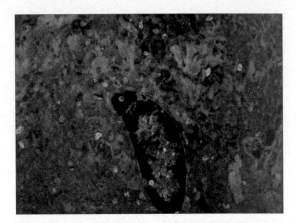

Figure 3. Human carotid plaque. Immunofluorescence showing dendritic cells, stained with a fluorescein isothiocyanate-labeled CD68 antibody, within an hemorrhagic plaque (red blood cells stained with a rhodamine-labeled antiglycophorin antibody). Nuclei were counterstained with DAPI (200×).

ders of atheromas. In this way, an immune response against endogenous modified antigens takes place, stimulating a further synthesis of cytokines and chemokines. This self-renewing process gives rise to chronic low-level inflammation, contributing to the development of complex atherosclerotic plaques, which occasionally ulcerate, often causing fatal clinical events.

Cytokines and plaque progression

T cells recognize modified autoantigens that are presented by antigen-presenting cells, such as macrophages or dendritic cells, inducing the release of chemokines, cytokines, and proteases that enhance the recruitment of monocytes and lymphocytes, thus promoting the progression of atherosclerotic lesions.[35] Th1 cells are responsible for cell-mediated immunity, and their activation is associated with the presence of proinflammatory cytokines, including IFN-γ, IL-2, and IL-22, that promote the development of the disease. Less abundant within atherosclerotic plaques, Th2 cells are responsible for humoral immunity and are characterized by the production of anti-inflammatory cytokine such as IL-4, IL-5, IL-10, and IL-13.[35,36] The original paradigm of Th1/Th2 turned out to be more complicated than initially stated. Two additional lineages of T cells are now known to be important regulators of immune responses, namely Th17 cells, involved in autoimmune diseases, and

T_{reg} cells, involved in suppressing pathogenic Th1 and Th2 responses.[37,38]

Unstable atherosclerotic plaques, vulnerable to rupture, usually contain an increased accumulation of inflammatory cells, particularly macrophages and T lymphocytes that secrete growth factors and cytokines.[39,40]

It is becoming clear that plaque progression to an unstable phenotype, as a prelude to thromboembolism following the rupture of the plaque, depends on the cell subtypes and on the specific mediators accumulating within the plaque. A number of studies that are now in progress aim to elucidate this issue, especially focusing on cell subtypes and molecules stimulating the thinning of the fibrous cap and its subsequent rupture. For several years we have been studying the mediators released by the immune system associated with the development of carotid atherosclerosis. Considering that pro- and anti-inflammatory balance is a major determinant of disease progression, we determined by multiparametric flow cytometry the expression of proinflammatory cytokines, including TNF-α, IFN-γ, IL-1β, IL-6, and IL-8, in peripheral blood lymphocytes and monocytes from patients with carotid atherosclerosis. Increased intracellular cytokines in patients' peripheral blood might be a warning signal indicating progressive atherosclerosis. Intracellular cytokine expression was significantly higher in patients undergoing carotid endarterectomy (CEA) who had stenosis ≥70% (TNF-α, IFN-γ, IL-1β, IL-6), with previous stroke (IFN-γ, IL-1β, IL-6, IL-8), and with amaurosis fugax (IFN-γ, IL-6) than in patients with a stable plaque without indication for CEA.[41,42]

The IL-12 family of cytokines is a key player in the regulation of T cell responses.[43] These responses are orchestrated by monocytes, macrophages, and dendritic cells that produce members of the IL-12 family of cytokines in response to different stimuli. IL-27 and IL-23 are two cytokines related to IL-12; these cytokines share homology at the subunit, receptor, and signaling levels. IL-12 is composed of p35 and p40 subunits, which, when combined form the bioactive IL-12p70. IL-23 is composed of the IL-12p40 subunit as well as the IL-23p19 subunit, which shares homology with IL-12p35. IL-27 is composed of EBI3 and p28. These three cytokines activate similar members of the JAK/STAT signaling pathways as a result of homology in their receptor components. Production of these cytokines by activated monocytes, macrophages, and dendritic cells results in the activation and differentiation of T cells. In spite of their similarity, each of these cytokines has specific roles in the regulation of immune responses. IL-12 is required for the induction of IFN-γ production, critical for the induction of Th1 cells; IL-27 has been shown to play a role in the induction of Th1 cells from naive T cells; whereas IL-23 has been demonstrated to play a key role in the induction and maintenance of the newly described Th17 cells.[43] IL-17A is also involved in the pathology of several autoimmune diseases, mainly via regulation of chemokine expression and leukocyte migration to the site of inflammation. There is an increasing body of evidence indicating an association between elevated levels of IL-17A and cardiovascular diseases.[44] Functional blockade of IL-17A reduces atherosclerotic lesion development and decreases plaque vulnerability, cellular infiltration, and tissue activation in apolipoprotein E-deficient mice.[45] Atheromas of symptomatic patients, characterized by a highly activated inflammatory milieu, showed significantly higher IL-17A expression levels, as well as markedly higher number of IL-17A–expressing cells.[46,47] Patients with acute coronary syndrome revealed a significant increase in peripheral Th17 number and Th17-related cytokines (IL-17, IL-6, IL-23), and a decrease in T_{reg} cell number.[48] The Th17/ T_{reg} cell balance controls inflammation and may be important in the pathogenesis of plaque destabilization.

Conclusions

Atherosclerotic plaque development involves multiple risk factors and chronic inflammatory processes. Knowledge of cellular and molecular players in the progression of atherosclerotic plaque has dramatically improved in the last two decades, by assessing the central role of innate and adaptive immunity cells and proinflammatory cytokines driving the evolution of the plaques. The elucidation of molecular pathways underlying thrombotic events at the level of the plaque should provide new important therapeutical tools to counteract main cardiovascular disease onset.

Conflicts of interest

The authors declare no conflicts of interest.

References

1. Ross, R. 1999. Atherosclerosis—an inflammatory disease. *N. Engl. J. Med.* **340:** 115–126.
2. Hansson, G.K. & A. Hermansson. 2011. The immune system in atherosclerosis. *Nat. Immunol.* **12:** 204–212.
3. Libby, P. *et al.* 2011. Progress and challenges in translating the biology of atherosclerosis. *Nature* **473:** 317–325.
4. Nakashima, Y. *et al.* 2007. Early human atherosclerosis: accumulation of lipid and proteoglycans in intimal thickenings followed by macrophage infiltration. *Arterioscler. Thromb. Vasc. Biol.* **27:** 1159–1165.
5. Little, P.J. *et al.* 2011. Cellular and cytokine-based inflammatory processes as novel therapeutic targets for the prevention and treatment of atherosclerosis. *Pharmacol. Ther.* **131:** 255–268.
6. González-Chávez, A. *et al.* 2011. Pathophysiological implications between chronic inflammation and the development of diabetes and obesity. *Cir. Cir.* **79:** 209–216.
7. Businaro, R. *et al.* 2005. Morphological analysis of cell subpopulations within carotid atherosclerotic plaques. *Ital. J. Anat. Embryol.* **110:** 109–115.
8. Gordon, S. & A. Mantovani. 2011. Diversity and plasticity of mononuclear phagocytes. *Eur. J. Immunol.* **41:** 2470–2472.
9. Woollard, K.J. & F. Geissmann. 2010. Monocytes in atherosclerosis: subsets and functions. *Nat. Rev. Cardiol.* **7:** 77–86.
10. Combadière, C. *et al.* 2008. Combined inhibition of CCL2, CX3CR1, and CCR5 abrogates Ly6C(hi) and Ly6C(lo) monocytosis and almost abolishes atherosclerosis in hypercholesterolemic mice. *Circulation* **117:** 1649–1657.
11. Moore, K.J. & I. Tabas. 2011. Macrophages in the pathogenesis of atherosclerosis. *Cell* **145:** 341–355.
12. Weber, C. & H. Noels. 2011. Atherosclerosis: current pathogenesis and therapeutic options. *Nat. Med.* **17:** 1410–1422.
13. Drechsler, M. *et al.* 2010. Hyperlipidemia-triggered neutrophilia promotes early atherosclerosis. *Circulation* **122:** 1837–1845.
14. Guasti, L. *et al.* 2011. Neutrophils and clinical outcomes in patients with acute coronary syndromes and/or cardiac revascularisation. A systematic review on more than 34,000 subjects. *Thromb. Haemost.* **106:** 591–599.
15. Guasti, L. *et al.* 2008. Prolonged statin-associated reduction in neutrophil reactive oxygen species and angiotensin II type 1 receptor expression: 1-year follow-up. *Eur. Heart J.* **29:** 1118–1126.
16. Baetta, R. & A. Corsini. 2010. Role of polymorphonuclear neutrophils in atherosclerosis: current state and future perspectives. *Atherosclerosis* **210:** 1–13.
17. Geissmann, F. *et al.* 2003. Blood monocytes consist of two principal subsets with distinct migratory properties. *Immunity* **19:** 71–82.
18. Gordon, S. & P.R. Taylor. 2005. Monocyte and macrophage heterogeneity. *Nat. Rev. Immunol.* **5:** 953–964.
19. Mantovani, A. *et al.* 2009. Macrophage diversity and polarization in atherosclerosis: a question of balance. *Arterioscler. Thromb. Vasc. Biol.* **29:** 1419–1423.
20. Pello, O.M. *et al.* 2011. A glimpse on the phenomenon of macrophage polarization during atherosclerosis. *Immunobiology* **216:** 1172–1176.
21. Khallou-Laschet, J. *et al.* 2010. Macrophage plasticity in experimental atherosclerosis. *PLoS One* **5:** e8852.
22. Lahoute, C. *et al.* 2011. Adaptive immunity in atherosclerosis: mechanisms and future therapeutic targets. *Nat. Rev. Cardiol.* **8:** 348–358.
23. Taleb, S. *et al.* 2010. Adaptive T cell immune responses and atherogenesis. *Curr. Opin. Pharmacol.* **10:** 197–202.
24. Shoenfeld, Y. *et al.* 2005. Accelerated atherosclerosis in autoimmune rheumatic diseases. *Circulation* **112:** 3337–3347.
25. Tsan, M.F. & B. Gao. 2004. Heat shock protein and innate immunity. *Cell Mol. Immunol.* **1:** 274–279.
26. Riganò, R. *et al.* 2007. Heat shock proteins and autoimmunity in patients with carotid atherosclerosis. *Ann. N.Y. Acad. Sci.* **1107:** 1–10.
27. Businaro, R. *et al.* 2009. Heat-shock protein 90: a novel autoantigen in human carotid atherosclerosis. *Atherosclerosis* **207:** 74–83.
28. Buttari, B. *et al.* 2011. Oxidized human beta2-glycoprotein I: its impact on innate immune cells. *Curr. Mol. Med.* **11:** 719–725.
29. Buttari, B. *et al.* 2011. Advanced glycation end products of human β_2 glycoprotein I modulate the maturation and function of DCs. *Blood* **117:** 6152–6161.
30. Profumo, E. *et al.* 2009. Oxidized haemoglobin as antigenic target of cell-mediated immune reactions in patients with carotid atherosclerosis. *Autoimmun. Rev.* **8:** 558–562.
31. Schaer, D.J. *et al.* 2006. CD163 is the macrophage scavenger receptor for native and chemically modified hemoglobins in the absence of haptoglobin. *Blood* **107:** 373–380.
32. Yunoki, K. *et al.* 2009. Enhanced expression of haemoglobin scavenger receptor in accumulated macrophages of culprit lesions in acute coronary syndromes. *Eur. Heart J.* **30:** 1844–1852.
33. Buttari, B. *et al.* 2007. Free hemoglobin: a dangerous signal for the immune system in patients with carotid atherosclerosis? *Ann. N.Y. Acad. Sci.* **1107:** 42–50.
34. Buttari, B. *et al.* 2011. Haemoglobin triggers chemotaxis of human monocyte-derived dendritic cells: possible role in atherosclerotic lesion instability. *Atherosclerosis* **215:** 316–322.
35. Ait-Oufella, H. *et al.* 2009. Cytokine network and T cell immunity in atherosclerosis. *Semin. Immunopathol.* **31:** 23–33.
36. Ait-Oufella, H. *et al.* 2011. Recent advances on the role of cytokines in atherosclerosis. *Arterioscler. Thromb. Vasc. Biol.* **31:** 969–979.
37. Park, H. *et al.* 2005. A distinct lineage of CD4 T cells regulates tissue inflammation by producing interleukin 17. *Nat. Immunol.* **6:** 1133–1141.
38. Sakaguchi, S. *et al.* 2008. Regulatory T cells and immune tolerance. *Cell* **133:** 775–787.
39. Frostegård, J. *et al.* 1999. Cytokine expression in advanced human atherosclerotic plaques: dominance of pro-inflammatory (Th1) and macrophage-stimulating cytokines. *Atherosclerosis* **145:** 33–43.

40. Tedgui, A. & Z. Mallat. 2006. Cytokines in atherosclerosis: pathogenic and regulatory pathways. *Physiol. Rev.* **86:** 515–581.

41. Profumo, E. *et al.* 2007. Intracellular expression of cytokines in peripheral blood from patients with atherosclerosis before and after carotid endarterectomy. *Atherosclerosis* **191:** 340–347.

42. Profumo, E. *et al.* 2008. Association of intracellular pro- and anti-inflammatory cytokines in peripheral blood with the clinical or ultrasound indications for carotid endarterectomy in patients with carotid atherosclerosis. *Clin. Exp. Immunol.* **152:** 120–126.

43. Gee, K. *et al.* 2009. The IL-12 family of cytokines in infection, inflammation and autoimmune disorders. *Inflamm. Allergy Drug Targets* **8:** 40–52.

44. Butcher, M. & E. Galkina. 2011. Current views on the functions of interleukin-17A-producing cells in atherosclerosis. *Thromb. Haemost.* **106:** 787–795.

45. Erbel, C. *et al.* 2009. Inhibition of IL-17A attenuates atherosclerotic lesion development in apoE-deficient mice. *J. Immunol.* **183:** 8167–8175.

46. Erbel, C. *et al.* 2011. Expression of IL-17A in human atherosclerotic lesions is associated with increased inflammation and plaque vulnerability. *Basic Res. Cardiol.* **106:** 125–134.

47. de Boer, O.J. *et al.* 2010. Differential expression of interleukin-17 family cytokines in intact and complicated human atherosclerotic plaques. *J. Pathol.* **220:** 499–508.

48. Cheng, X. *et al.* 2008. The Th17/Treg imbalance in patients with acute coronary syndrome. *Clin. Immunol.* **127:** 89–97.